RAND NATIONAL DEFENSE RESEARCH INSTITUTE

Retention, Incentives, and DoD Experience Under the 40-Year Military Pay Table

Beth J. Asch, James Hosek, Jennifer Kavanagh, Michael G. Mattock

Prepared for the Office of the Secretary of Defense for Personnel and Readiness

For more information on this publication, visit www.rand.org/t/rr1209

Library of Congress Cataloging-in-Publication Data
ISBN: 978-0-8330-9175-8

Published by the RAND Corporation, Santa Monica, Calif.

Preface

The Senate Armed Services Committee (SASC) report accompanying the fiscal year 2015 National Defense Authorization Act (NDAA) directed the Secretary of Defense to review the military's pay tables, focusing on whether the 40-year pay table is still justified as a retention tool. Congress extended the pay tables to 40 years as part of the fiscal year 2007 NDAA to provide an incentive for the most-experienced members to continue to serve. But the SASC report argued that the military is now drawing down and asked whether it is useful to continue the 40-year table from a retention standpoint, or return to the 30-year table.

In preparing its report to the SASC, the U.S. Department of Defense asked the RAND Corporation to provide analytic support, and this report documents RAND's research. It should be of interest to policymakers and researchers concerned with military compensation and the relationship between the structure of the military pay table and the retention of enlisted and officer personnel.

This research was sponsored by the Office of the Under Secretary of Defense for Personnel and Readiness and conducted within the Forces and Resources Policy Center of the RAND National Defense Research Institute, a federally funded research and development center sponsored by the Office of the Secretary of Defense, the Joint Staff, the Unified Combatant Commands, the Navy, the Marine Corps, the defense agencies, and the defense Intelligence Community.

For more information on the Forces and Resources Policy Center, see www.rand.org/nsrd/ndri/centers/frp.html or contact the director (contact information is provided on the web page).

Contents

Figures and Table

Figures

Table

Summary

The military basic pay table is the foundation of military compensation for currently serving members. Until 2007, basic pay increases associated with additional years of service—called longevity increases—occurred only up to year of service (YOS) 26 in the basic pay table. In 2007, this so-called 30-year pay table was replaced with a 40-year table that added longevity increases beyond YOS 26 and additional increases beyond YOS 30 to the highest-ranked personnel. Specifically, the 40-year table added longevity increases for officers in grades O-6 and above, warrant officers in grades W-4 and W-5, and enlisted personnel in grades E-8 and above. The major objective of the 40-year table was to provide incentives for longer careers, especially to the most senior-ranked officers, but also to more-junior officers aspiring to higher-ranked positions.[1]

As part of the fiscal year (FY) 2015 National Defense Authorization Act (NDAA), the Senate Armed Services Committee (SASC) requested that the U.S. Department of Defense (DoD) review the 40-year pay table and report on whether it is necessary for retaining experienced personnel or whether such retention could be equally achieved with a 30-year pay table (SASC, 2014). The Office of Compensation within the Office of the Under Secretary of Defense for Personnel and Readiness asked the RAND Corporation to provide information to support its review. This document summarizes the analysis we conducted to meet this request.

Approach

Our study took a multimethod approach. We obtained background information to understand both the historical context of the 2007 legislation and the specific effects of that legislation on the basic pay of military members by grade. We reviewed previous literature on the theoretical underpinnings of the military compensation structure by grade. In addition, we used Defense Manpower Data Center pay file data on active-duty personnel by grade, years of service, and service to tabulate the number of personnel serving with more than 30 years of service before and after 2007, specifically between 2000 and 2014. These tabulations were specifically requested by the SASC as part of the pay table review and provided information on the degree to which the services' use of personnel with more than 30 years of service has changed since 2007.

[1] In addition, the 2007 legislation included targeted pay raises to warrant officers and mid-career enlisted personnel in grades E-5 to E-7. These changes addressed a recommendation of the Ninth Quadrennial Review of Compensation, which called for improved pay comparability between basic pay at these grades and pay for workers with comparable education and experience in the general economy.

We also used qualitative and quantitative methods to assess the retention effects of the 40-year versus 30-year pay table. First, we conducted semistructured interviews in May and June 2015 with senior military and civilian personnel within DoD and across the services who have experience and knowledge relevant to military compensation, retention, and personnel management. The purpose of the interviews was to get experts' perspectives on the performance of the 40-year versus 30-year table in meeting retention goals, possible changes in personnel management as a result of the change to the 40-year table, and whether returning to the 30-year table would require changes in other policies, such as the use of special and incentive pays. Second, we used the RAND Dynamic Retention Model (DRM) to

- simulate the effect on retention of the compensation changes that occurred in 2007
- simulate the effect on retention of reverting to a 30-year pay table, in terms of longevity increases beyond YOS 26[2]
- assess whether and how much special and incentive pay would be needed to sustain retention under a 30-year table
- compute the additional cost of operating under a 30-year table versus a 40-year table while still sustaining retention (information on cost was another element of the review requested by the SASC).

Key Findings

The Services Made Greater Use of Senior Personnel Since 2007

Our tabulations revealed that the number of active-duty personnel with more than 30 years of service increased by 58 percent between 2007 and 2014, from 4,175 to 6,583. However, the greater percentage increase was not among general and flag officers, the group that represented the impetus for changing to the 40-year table. That group—specifically officers in O-7 to O-10—increased by only 6.5 percent over that period. Instead, the greatest increase was among enlisted personnel, particularly E-9s; the number of enlisted personnel with more than 30 years of service more than doubled between 2007 and 2014, from 809 to 2,029. This trend actually began before 2007, implying that the increase in senior enlisted personnel after 2007 was part of an ongoing trend. Another major source of growth was among senior field grade officers in O-4 to O-6, increasing by nearly 50 percent. It is likely that most of these were people with prior enlisted service, because O-6s are not permitted to serve with more than 30 years of commissioned service unless they have a waiver. Although the effect for the most highly ranked officers was smaller, the direction is consistent with the objective of the legislation, to increase the retention of this group of officers.

While these tabulations suggest that the 2007 pay table change had a substantial effect, we cannot attribute a causal effect. Other factors may have caused the increase, including the other compensation changes that occurred in 2007. Perhaps most importantly, the period before and after 2007 was one of high demand for military forces because of the wars in Afghanistan and Iraq. Requirements for personnel increased, and the services used emergency authorities

[2] Throughout the report, when we refer to reverting to the 30-year table, we mean eliminating longevity increases beyond YOS 26 but keeping the same structure of pay for those with fewer than 26 years of service. Thus, the targeted mid-career pay raises that were provided in 2007 are retained.

that allowed them to exceed certain caps on the number of senior personnel. Thus, the increase in the number of senior personnel serving after 2007 could be due to rising requirements. Of course, the 40-year pay table may have provided the additional incentives needed to enable the services to meet this higher requirement, but we cannot discern from the trend data alone the extent to which that was the case.

We also considered five-year continuation rates to YOS 30 among those with 26 years of service, as well as two-year continuation rates to YOS 32 among those with 30 years of service. While continuation rates varied considerably between 2000 and 2011, we did not observe a large increase in rates overall after 2007. The lack of a marked change in overall rates of continuation after 2007, despite the increase in the numbers of personnel with more than 30 years of service, is consistent with the argument that requirements for senior personnel increased after 2007 and that these requirements were filled by senior personnel in very specific groups, such as officers with prior enlisted experience who stayed for an extra assignment, senior enlisted and warrant officers who similarly were retained to fill specific jobs, and recalled retirees who returned to support the increased pace of deployment.

Interviews Indicated That Both the 30-Year and 40-Year Tables Performed Well, but Reverting to the 30-Year Table Would Be Undesirable

The experts we interviewed said that both the 40-year pay table and its predecessor, the 30-year table, have proven satisfactory overall in providing the services with the retention profiles needed by years of service, as well as with personnel pools of sufficient quality and size from which to select senior leaders. In fact, none of the experts felt that the 40-year pay table was necessary to successfully retain the most-experienced personnel, as long as no other 2007 change in compensation is reversed. The reasons given varied, including the observations that senior military personnel are highly selected and serve because of their strong commitment to the military (not because of longevity increases) and that the services actively manage their most-senior leaders and use personnel management tools to ensure that these positions are filled to meet requirements.

Nearly all of the experts also agreed that returning to the 30-year table was a bad idea, again for a variety of reasons. Many cited the adverse effect of such a change on the morale of both military personnel and their spouses. Others said that the reversal would contribute to an overall perception that military compensation is unstable and less valuable, coming on top of recent changes to the military retirement system. Some argued that the 40-year table provides additional flexibility to achieve longer careers by offering more-generous longevity increases in the most-senior grades. While more-junior personnel could be promoted more quickly to fill these positions, it was argued that it is preferable to retain people who already have the needed experience, especially because the services could induce mandatory separation for these people if required. Other experts mentioned the increased requirements for senior personnel, not only as a result of operations in Afghanistan and Iraq but also because the nature of military service might be changing in a way that puts greater emphasis on technical skills and experience. While few interviewees thought that going back to a 30-year table would have any effect on retaining senior officers, some expressed concern that it could have a negative effect on enlisted personnel and warrant officers, because these personnel earn less, resulting in a larger effect on their earnings. In sum, virtually all of the experts supported keeping the 40-year pay table.

Reverting to the 30-Year Table Would Require Additional Special Pays Targeted to Senior Personnel to Sustain Retention

We used the DRM to assess whether retention could be sustained under the 30-year pay table, relative to the 40-year pay table, and, if not, how large of a special pay would be needed and at what cost. We conducted the analysis for the active component of all services and describe the results for the Army in this report. We found that reverting to a 30-year pay table would adversely affect active-component retention, especially among personnel with more than 30 years of service, but also among those with more than 20 years. We found that a special pay given at YOS 30 of between $87,900 and $99,600 for officers and between $37,400 and $58,200 for enlisted personnel (in 2015 dollars) would restore retention relative to the 40-year pay table, assuming no other change in compensation. Using the DRM, we estimate that this would produce a cost savings for the active component of about $1.2 billion. This estimate is based on the DRM simulated retention profiles; DoD (including the DoD Office of the Actuary, which uses the retention profile for recent years and its own costing methodology) would make the official computation of cost savings for any budget planning purposes.

Thus, if the 30-year table were brought back, a special pay targeted to senior personnel would be needed and would be effective. Our interviews indicated that while the services have the legal authority to use assignment incentive pay to retain senior personnel, current DoD policy does not permit them to do so. The DRM results suggest that if the 30-year table were restored, DoD should revisit current instructions and policies on using assignment incentive pay.

Economic Theory Justifies Larger Increases in Military Compensation at Higher Grades

Economic theory provides a foundation for considering how military compensation should be structured between grades to meet the services' manpower requirements—for example, to attract and retain personnel, motivate them, sort the most talented and induce them to stay and seek the most-senior positions, and eventually separate them. The theory, laid out in various literature, accounts both for the unique aspects of military service, including the lack of lateral entry in the active force, and for individual decisionmaking with respect to retention and effort.

A key insight from the theory is that the military compensation structure should be *skewed*, meaning that the gaps in compensation between ranks should be sequenced so that the gaps increase with grade. The growing differential in compensation between grades means that the payoff for being promoted increases with rank. Such a structure is necessary to offset factors that tend to discourage effort and the retention of the most-talented personnel. These factors include the difficultly of measuring and assessing performance in the upper ranks, the greater homogeneity of the talent pool in the upper ranks, and the declining number of future promotions (and hence future increases in compensation) at higher ranks. Other factors decrease the size of the compensation gaps required, including the higher value that members attach to the nonpecuniary aspects of serving in higher ranks and the stronger taste for or commitment to the military in higher grades.

The theory does not provide guidance on how much skewness is required, but it offers little support for ending those incentives at YOS 30, especially if the larger number of personnel who serve with more than 30 years of service is expected to continue. While much of the skewness of the current compensation system occurs through the military retirement system for the most-senior officers in grades O-9 and O-10 rather than through basic pay increases, basic pay increases also increase retirement benefits. Furthermore, basic pay increases in the pay table do affect skewness for senior enlisted personnel.

Conclusions

A 30-year table could be as effective at sustaining retention as a 40-year table, as long as the services had adequate special pay to manage retention of senior personnel. In the absence of such special pay, our analysis using the DRM shows that the 30-year table would hurt retention. Still, even when utilizing special pay with the 30-year table, there are several reasons why continuing the 40-year table is preferred. It performs well and many argue that it improves readiness and flexible personnel management. Reverting to the 30-year table could adversely affect morale and perceptions about the stability and value of military compensation overall, especially in the context of recent changes to the military retirement system. Reverting to a 30-year table will affect 58 percent more people than it did prior to 2007, potentially aggravating these perceptions. Finally, the cost of keeping the 40-year table is relatively small, resulting in a 1.1-percent change in active-component personnel cost relative to using a 30-year table with special pay to sustain retention.

Acknowledgments

We are indebted to the military personnel management experts who participated in the interviews we conducted. Their input enriches our study. We are grateful to Don Svendsen in the Office of Compensation within the Office of the Under Secretary of Defense for Personnel and Readiness. Svendsen served as project monitor, provided background material, and arranged interviews within the project's tight timeline. We are also grateful to Patricia Mulcahy, also in the Office of Compensation, and Jerilyn Busch, director of that office, who provided input and guidance for our analysis. At RAND, we wish to thank Arthur Bullock, who created data files and ran the retention tabulations in Chapter Five. Finally, we greatly appreciate the input of the two reviewers of an earlier version of this report, Amalia Miller and Saul Pleeter. The final report benefited from their comments.

Abbreviations

DoD	U.S. Department of Defense
DRM	Dynamic Retention Model
FY	fiscal year
GOFO	general officer/flag officer
NDAA	National Defense Authorization Act
OSD	Office of the Secretary of Defense
SASC	Senate Armed Services Committee
YOS	year of service

CHAPTER ONE

Introduction

The centerpiece of military compensation for currently serving members is basic pay, which depends on a service member's pay grade and years of service. While not the only form of compensation members receive,[1] basic pay accounts for approximately 60 percent of current compensation for active-component members and is used as the basis for computing the military retirement benefit. It increases within a grade as years of service increase and between grades as a result of promotion. There are separate basic pay tables for commissioned officers, warrant officers, and enlisted personnel. Until 2007, longevity increases stopped at year of service (YOS) 26. That is, members serving beyond 26 years no longer received basic pay increases as a result of additional seniority. This pay structure has been termed the *30-year basic pay table.*

The National Defense Authorization Act (NDAA) for fiscal year (FY) 2007 extended the pay table for active- and reserve-component personnel from 30 to 40 years of service. The 40-year table, effective April 2007, added longevity increases after YOS 26 for officers in O-6 and above, warrant officers in W-4 and W-5, and enlisted members in E-8 and E-9. The intent of this change was to provide an incentive for the most-experienced members to continue to serve and to reward such service. The same legislation raised the cap on basic pay for general and flag officers (O-7 to O-10) as of January 2007 from Executive Level III to Executive Level II, and it eliminated the cap on basic pay for the purpose of computing retired pay. Under the 30-year pay table, the maximum multiplier was 75 percent at YOS 30. Thus, under the 2007 change, the multiplier is 100 percent of basic pay at YOS 40. In sum, the following four changes for senior personnel occurred in 2007:[2]

- Move from a 30-year to 40-year pay table.
- Increase the cap on basic pay for senior personnel from Executive Level III to Executive Level II.
- Remove the Executive Level cap on basic pay for the purpose of computing retired pay.[3]
- Remove the cap on years of service for computing the retirement benefit multiplier, previously 30 years of service (or 75-percent multiplier).

[1] Members also receive a basic allowance for housing, a basic allowance for subsistence, and a tax advantage (because allowances are nontaxable).

[2] The legislation also included targeted pay raises for mid-career enlisted and warrant officer personnel, but those changes are not the focus of our analysis.

[3] This cap was reinstated in the FY 2015 NDAA.

The Senate Armed Services Committee (SASC) report accompanying S. 2410 (the Senate version of the FY 2015 NDAA) directed the Secretary of Defense to review the military pay tables, focusing on whether the 40-year pay table is still justified as a retention tool (SASC, 2014). The SASC report argued that the military is now drawing down and asked whether it is useful to continue the 40-year table from a retention standpoint, or whether the retention of experienced personnel who would otherwise be difficult to retain could be achieved with a 30-year pay table. SASC also asked the Secretary of Defense report to provide a description of how many remained on active duty past YOS 30 since 2007, a breakdown by grade, the additional costs since 2007 of operating under the 40-year rather than the 30-year table, and an assessment of how longevity pay increases beyond YOS 30 affect retention.

To support its review in response to the SASC request, the director of the Office of Compensation within the Office of the Under Secretary of Defense for Personnel and Readiness requested analysis from the RAND Corporation. This report summarizes the analysis that we provided. The objectives of the research were to analyze the pay table and, importantly, to provide information on the effect on retention of the 40-year table versus returning to a 30-year table.[4]

We begin in Chapter Two with a history of the 40-year table and describe the effects on basic pay and retired pay of the legislative changes in 2007. In Chapter Three, we review the defense manpower literature on the structure of military compensation. The purpose of the review is to gain insight into how the level of military compensation in the senior ranks relative to the levels in mid-career and junior ranks affects the ability of military compensation to attract and retain talented personnel, induce them to stay and seek higher-ranked positions, and eventually induce them to leave at the end of their careers.

Chapter Four summarizes the themes emerging from our interviews of military personnel experts, civilian and military, who have insight into compensation and the management of senior military personnel; these interviews provided qualitative assessments of the effectiveness of the 40-year versus 30-year pay table. The purpose of the interviews was to obtain input on the operation, benefits, and obstacles associated with the 40-year versus 30-year table, especially input that may be difficult to measure or evaluate quantitatively. Next, in response to the SASC request, Chapter Five presents tabulations on retention and personnel strength trends among military personnel in YOS 20–40 between 2000 and 2014, covering a period before and after the 2007 pay table change. These tabulations provide context on the retention of more-senior personnel and how retention changed over the past dozen or so years.

Chapter Six summarizes the key findings from simulations conducted with RAND's Dynamic Retention Model (DRM). The simulations show voluntary retention behavior and costs under a return to the 30-year table while controlling for features of the 2007 legislation that might not be changed, including imposing the Executive Level II cap on basic pay, allowing years of service beyond 30 to be counted in determining the retirement benefit, and reinstating the Executive Level II cap on basic pay in determining the retirement benefit. We compare the results of these simulations with the retention and cost of the 40-year pay table and its provisions. We also ran simulations to show the effect of the provisions individually, such as reimposing the Executive Level II cap when computing the retirement benefit. Further,

[4] Throughout the report, when we refer to reverting to the 30-year table, we mean eliminating longevity increases beyond YOS 26 but keeping the same structure of pay for those with fewer than 26 years of service. Thus, the targeted mid-career pay raises that were provided in 2007 are retained.

we conducted simulations to determine the additional special and incentive pays (and associated costs) that might be needed to sustain retention under the 30-year table. The DRM is an econometric model of individual retention behavior in the military, estimated with 20 years of longitudinal data. We used the estimated model to simulate how changes in the level and structure of military compensation affect retention and cost. As described later, the DRM has been documented extensively and used for analyses of the retention and cost effects of other changes to military compensation, including pay raises and retirement reform proposals.

Thus, our approach draws upon policy context, peer-reviewed literature, expert knowledge, administrative data for retention tabulations, and advanced econometric methods for simulating policy to provide a basis for assessing the effect of introducing the 40-year pay table in 2007 and the possibility of returning from the 40-year pay table to a 30-year pay table. We present our conclusions in Chapter Seven.

CHAPTER TWO

Background on the Military Pay Table and the Move to the 40-Year Table

Compensation Under the 30-Year Pay Table

The FY 2007 NDAA replaced the 30-year military pay table with a 40-year pay table that increased the amount of compensation earned by the most-senior officers and senior enlisted personnel who had served in the military for longer than 30 years. Under the pre-2007 system, a 30-year pay table included pay raises by rank and years of service only through 26 years of service. The fact that no pay raises were scheduled past 26 years of service served as a disincentive for members to serve much longer. In addition, under the pre-2007 system, service members received no retirement credit for years of service beyond 30. This meant that when the multiplier of 2.5 percent of basic pay per year was applied to calculate retired pay, retirement was capped at 75 percent of basic pay (Henning, 2008), another disincentive for careers longer than 30 years. Thus, senior personnel did not earn additional pay raises after 26 years and also accrued no additional retirement earnings after 30 years. Because one year of retirement benefits is forgone for each year served past year 30, the absence of any increase in pay and retirement benefits after 30 years of service also meant there was no increase in compensation of any form to offset this loss.

Impetus for a Change

A primary impetus for the change to the 40-year pay table came from then–Secretary of Defense Donald Rumsfeld. Rumsfeld believed that the 30-year pay table and pre-2007 compensation system were insufficient to retain the experienced, high-quality personnel the military needed to succeed. In the case of officers, most generals reach the one-star rank at about the same point, 26 or 27 years of service, at which the longevity increases in the 30-year pay table and the associated incentives to continue military service came to an end. As Rumsfeld argued, the military often invested in and trained individuals to reach the general officer/flag officer (GOFO) level, then lost them after only one GOFO assignment. Rumsfeld compared this situation with the private sector, where senior executives are paid millions of dollars to secure the highest-quality talent and compensate them to handle high-level responsibilities critical to the health of the enterprise. The absence of basic pay increases after 26 years of service and retirement benefit increases after 30 years also provided senior enlisted personnel with no added incentive to stay in the military. The loss of highly skilled enlisted personnel became increasingly worrisome as military occupations grew increasingly technically and technologically advanced. Personnel in career fields where technical expertise and experience grew from a long military career were particularly hard to replace. Based on these observations and his experience in Washington and corporate America, Rumsfeld pushed for an increase in compensation for these individuals and a pay table that would incentivize service past 30 years.

Consistent with his concerns, a 2004 RAND report, *Aligning the Stars: Improvements to General and Flag Officer Management,* found that the military was not getting maximum benefit from its most-senior officers. Although officers were promoted to the highest rank (O-10) at about the same age as civilian counterparts became CEOs, most CEOs served for an average of 8.5 years and less than one-third retired before age 60, while O-10s served for about 3.5 years and almost 90 percent retired voluntarily before reaching age 60 (Harrell et al., 2004). The 9th Quadrennial Review of Military Compensation offered a related observation about military pay, especially that of mid-grade enlisted personnel and warrant officers. The report found a substantial pay gap between military and civilian pay that it argued would limit the services' ability to recruit and retain the high-quality personnel needed to complete their missions. The report also recommended that the compensation system be changed to close this pay gap to ensure that the military could recruit and keep the number and type of personnel it needed to be effective.

Military Compensation Under the 40-Year Pay Table

Responding to these pressures, the Office of Military Personnel Policy in the Office of the Secretary of Defense (OSD) developed a proposal to reform basic pay. The proposal included a 40-year pay table, across-the-board raises of at least 2.2 percent for all military grades, targeted increases for enlisted personnel at grades E-5 to E-7 and warrant officers, and changes to the executive pay cap to raise the pay of the most-senior officers. The proposal cited a number of motivations for these changes—for example, "to ensure that the uniformed services can recruit and retain a force of sufficient numbers and quality to support the military strategic and operational plans of this nation, military compensation must be adequate" (OSD, 2006). The proposal also noted, "given the changes in the technology of warfare, the compensation structure must change to accommodate longer career lengths" (OSD, 2006). Finally, the proposal was intended to improve equity and fairness by compensating senior personnel in a manner that would be considered comparable to private-sector practices.[1]

The targeted raises and the 40-year pay table responded to these concerns. The targeted increases for mid-grade enlisted personnel and warrant officers addressed the pay gap between military personnel and civilians that the 9th Quadrennial Review of Military Compensation had identified. The revisions to basic pay brought Regular Military Compensation to the 70th percentile of civilian pay when comparing education and experience. By adding longevity-based raises for personnel with more than 30 years of service, the longer pay table contained incentives for senior officers and enlisted personnel to stay in the military and made a career in the military more appealing (Pleeter, 2006; Chu, 2006). Specifically, the 40-year pay table added raises for officers in O-6 and above, warrant officers in W-4 and W-5, and enlisted personnel in E-8 and E-9. O-8s received pay raises at YOS 30 and 34. E-9s, W-5s, O-9s, and O-10s saw three raises, at YOS 30, 34, and 38. Congress made an additional change in the 2007 NDAA

---

[1] As mentioned, there were some targeted pay increases for mid-career personnel also for purposes of pay comparability.

that sharply increased pay for senior officers. Before 2007, basic pay for generals and admirals could not exceed Executive Level III pay for federal civilians, but starting in 2007, the cap was raised to Executive Level II pay. This pay increase provided further incentive for longer careers to those officers and lower-ranked officers aspiring to the highest ranks.

Some resisted the proposed reform of basic pay mainly because of the additional cost that would be incurred (Philpott, 2014). Some estimates suggested that the changes would increase the cost of providing military manpower by about $263 million per year (Pleeter, 2006). Other stakeholders felt that targeted increases for E-5 to E-7 personnel and warrant officers were not the best use of limited resources, arguing instead to target increases at groups with more-significant retention challenges, such as first-term personnel. Some opponents felt that the 40-year pay table would not have the desired effect of incentivizing longer careers among highly trained senior officers and enlisted personnel and might have unintended consequences, including reduced promotion opportunities and a resulting increase in separation of promising junior personnel when senior officers and enlisted remained in their positions for longer periods. Instead, some stakeholders suggested using individualized incentives to keep specific senior personnel who had the desired skills and experience.[2]

Despite the resistance, the changes to basic pay were written into the FY 2007 NDAA. The across-the-board 2.2-percent raise took effect on January 1, 2007, and the 40-year pay table took effect on April 1, 2007 (Pub. L. 109-364, 2006). Notably, the change to the 40-year pay table did not affect the high-year tenure rules that require retirement for officers and enlisted personnel based on grade and years of service if they have not been promoted.

Figures 2.1 through 2.4 illustrate the promotion-related and longevity increases in basic pay for officers, warrant officers, and enlisted personnel in January 2007 under the 30-year pay table and in April 2007 under the 40-year pay table. Again, the January 2007 basic pay amounts include the 2.2-percent across-the-board raise, so a comparison of the January and April 2007 basic pay amounts shows the impact of the targeted increases. Because the increases largely affect only the highest ranks, a relatively small number of personnel receive the increases, although the increases also provide an incentive to personnel at lower grades.

Figure 2.1 shows basic pay for officers in O-7 to O-10. Because the pay of senior officers is capped at Executive Level II pay—shown as "Max" in Figure 2.1—only O-8s realized basic pay increases after YOS 30, while both O-8s and O-9s received pay increases after YOS 28. Figure 2.2 shows basic pay for officers in O-4 through O-6. Although one might expect officers serving beyond YOS 30 to be O-7 and above, the tabulations in Chapter Five reveal an increase in O-5s and O-6s after 2007. Basic pay does not increase after 30 years of service for any of these grades, although O-6s do get one additional longevity increase at YOS 28 that was not included in the 30-year pay table. Figure 2.3 shows basic pay for warrant officers. Only W-5s receive raises after 30 years of service, although W-4s get an additional increase at 28 years that was not included in the shorter 30-year pay table. Chapter Five will also show an increase in W-5s (and a decrease in W-4s). Finally, for enlisted personnel (Figure 2.4), we see a steady increase in basic pay for E-9s after YOS 30, and E-8s receive a slight increase at YOS 28. Other enlisted personnel are not affected by the pay table change. Figures 2.3 and 2.4 also show basic pay at lower grades for completeness.

[2] Navy comments on "A Proposal to Increase Basic Pay Rates for Fiscal Year 2007," April 5, 2006, provided to RAND Corporation by OSD.

Figure 2.1
Basic Pay, Officers, O-7 to O-10

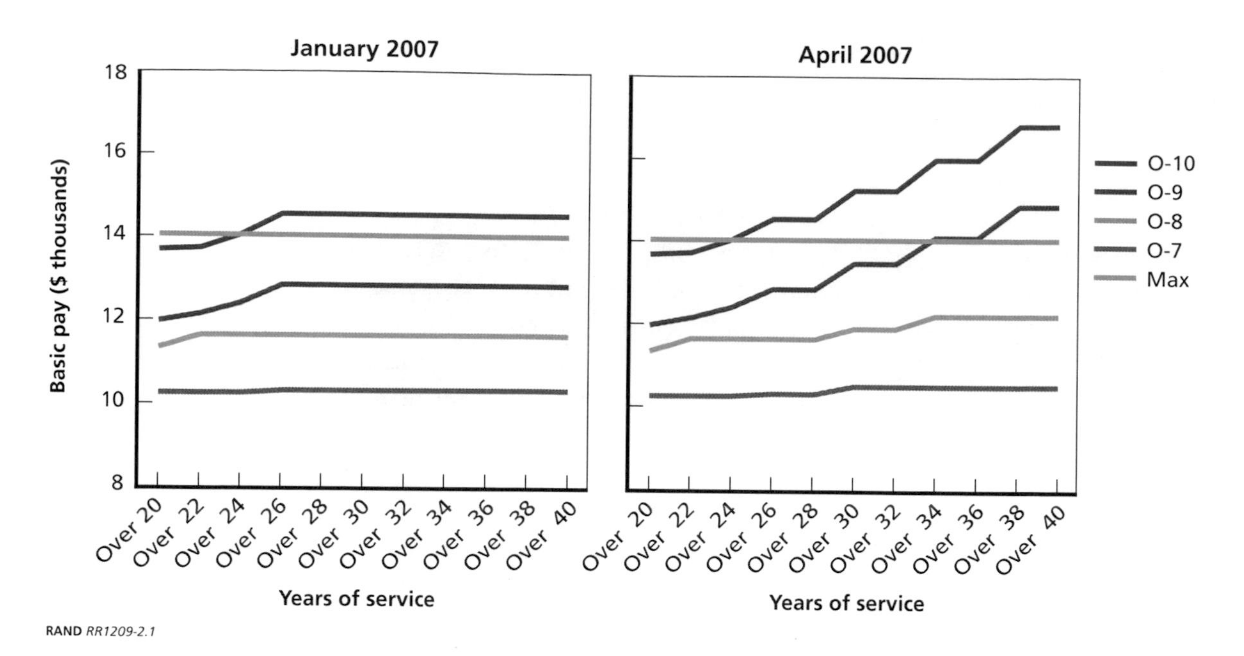

RAND *RR1209-2.1*

Figure 2.2
Basic Pay, Officers, O-4 to O-6

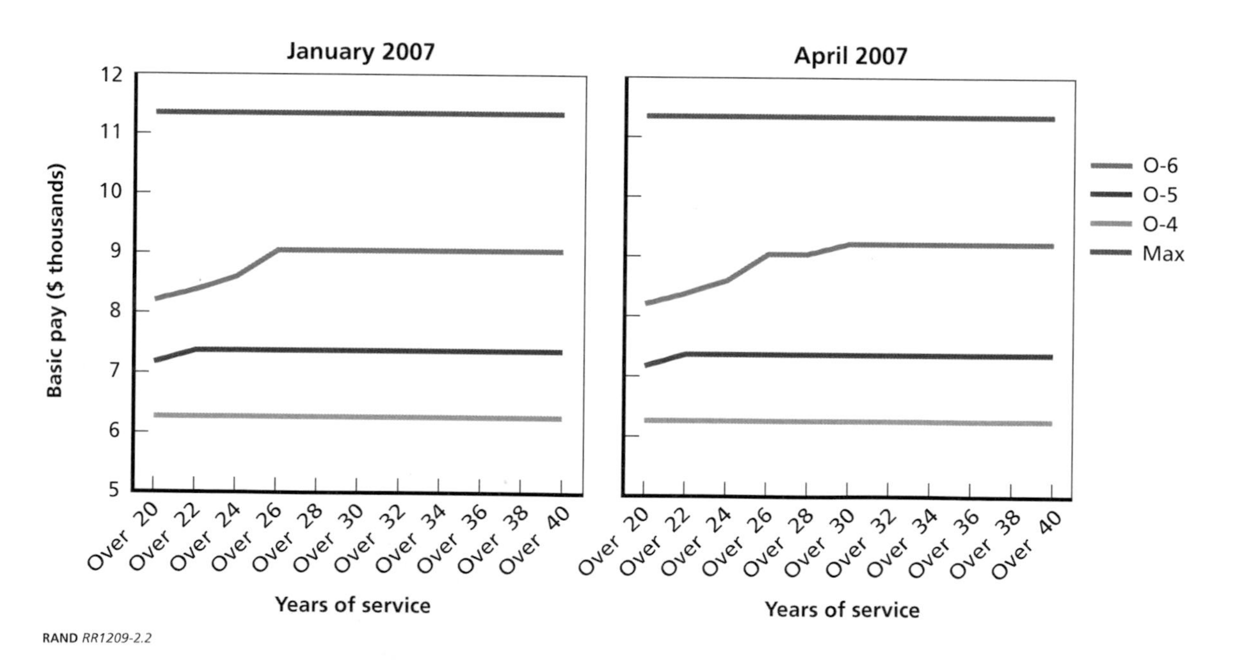

RAND *RR1209-2.2*

Figure 2.3
Basic Pay, Warrant Officers, W-1 to W-5

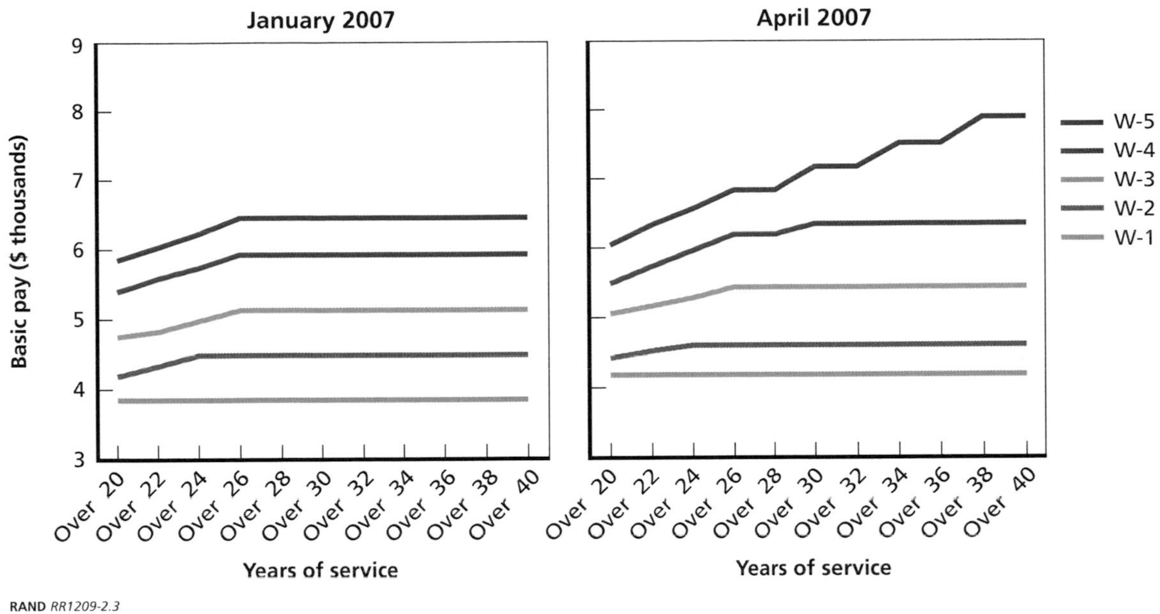

RAND *RR1209-2.3*

Figure 2.4
Basic Pay, Enlisted Personnel, E-5 to E-9

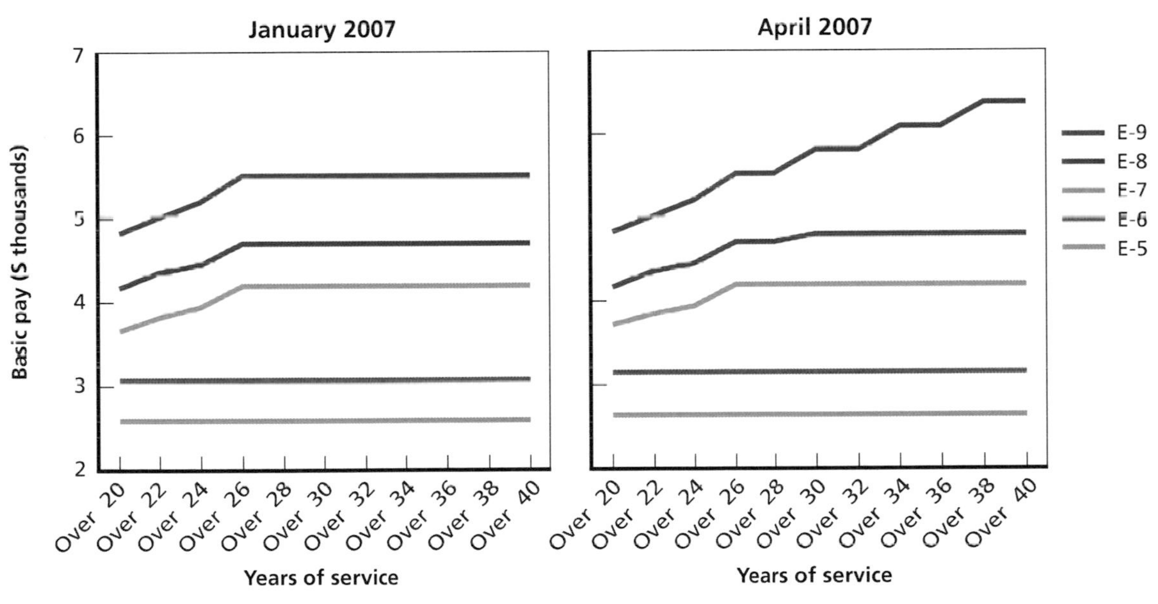

RAND *RR1209-2.4*

As mentioned, Congress made two additional changes in the 2007 NDAA that affected compensation for senior officers. These changes were not part of the 40-year pay table per se but were critically important for increasing compensation. First, the 30-year pay cap used in the calculation of retirement benefits before 2007 was revised so that service members who stayed longer than 30 years could continue to earn 2.5 percent of basic pay toward retirement all the way to 40 years, where annuities reach 100 percent of basic pay.[3] Second, Congress lifted the Executive Level pay cap completely when calculating retired pay of GOFOs. The latter change affected only a small population of the most-senior officers, but counting years of service after 30 in computing retirement benefits resulted in a significant increase in the compensation for anyone serving more than 30 years and provided a continuing incremental incentive for service even at pay grades where the post-service increases in basic pay from longevity were small.

Changes to the retirement system had a large effect on the retirement benefit of the highest-ranked officers. An O-10 with 38 years of service in January 2006 (for simplicity, assuming the retirement benefit was based on final pay) would retire with a monthly pay of 75 percent of the capped value of $12,666, or $113,994 per year. Raising the cap to $14,000 increased the annual retired pay to $126,000, or by approximately 10.5 percent. Basing retired pay on a multiplier of 95 percent (38 × 0.025) increased retired pay to $159,600, a 40-percent increase over the previous system of 75 percent of capped pay, even considering the cap of $14,000 per month. Finally, using the step on the April 2007 pay table for 38-plus years of service ($16,795.50; see Figure 2.1) with a multiplier of 95 percent yields annual retired pay of $191,469, a 68-percent increase.[4] The added longevity increases were insignificant compared with the changes in retirement to the O-10s and O-9s.

Similar calculations for an O-6, W-5, and E-9 with 38 years of service produce a 2006 retirement benefit of $79,572, $56,800, and $48,546, respectively, compared with an April 2007 benefit of $105,066, $90,247, and $72,754.

For the longest-serving, highest-rank officers, retirement benefits could be higher than compensation while serving in the military, given that their basic pay was capped at Executive Level II. Some suggested that this would provide a negative incentive to remain in the service, but no evidence was found to support this claim (Vanden Brook, 2012).

Recent Changes to Compensation

Under increasing budget pressures, Congress made recent changes affecting the compensation of senior officers. First, GOFOs did not receive the 1-percent pay raise in 2015 that was provided to the rest of the force. Second, the FY 2015 NDAA restored the Executive Level II cap on retired pay calculations. However, Congress included a provision to soften the effect of this change: It applies only to years served after 2014. It therefore does not affect those who retired before 2014 (Pub. L. 113-291, 2014). Further, for officers in GOFO grades serving after 2014 and who served in 2014 and before, their retirement benefit computation is calculated two ways, and their benefit is based on the higher of the two. One way uses their most recent basic pay, which might have increased if they were recently promoted, but is subject to the Level II

[3] More precisely, the basic pay used to calculate the retirement earnings was based on the member's highest 36 months of basic pay.

[4] We thank Saul Pleeter for this example.

cap for years 2015 onward when computing the retirement benefit. The other way uses basic pay from pre-2015 years of service and years of service as of December 31, 2014. Although years of service will be less and basic pay might be lower, it is not subject to the Level II cap when computing the benefit.[5] Finally, some legislators floated the idea of returning to a 30-year pay table—a topic leading to the present research.

[5] The relevant statute is Section 622 of the 2015 NDAA (Pub. L. 113-291, 2014), which in effect states that basic pay capped at Executive Level II should be used in computing the retirement benefit of officers retiring after December 31, 2014, unless the alternative method results in a higher benefit. Under the alternative method, "the amount of an officer's retired pay base shall be determined by using the rate of basic pay provided as of December 31, 2014, for that officer's grade as of that date for purposes of basic pay, with that officer's years of service creditable as of that date for purposes of basic pay."

CHAPTER THREE

Insights from the Literature on the Structure of Military Compensation

The SASC asked whether the 40-year pay table is necessary to retain the most-experienced personnel. This is one of a broader set of questions about how military compensation should be structured across ranks, given that the rank structure is hierarchical, the military promotes from within, the military does not permit lateral entry, and retention is voluntary so compensation must be high enough to induce talented junior and mid-ranked members to choose to stay and seek advancement to upper ranks.

A substantial literature has evolved in economics and management on how compensation should be structured across ranks in a large hierarchical organization, such as the military (Lazear and Oyer, 2012; Lazear and Rosen, 1981; Rosen, 1986; Malcomson, 1984). Rosen (1992) was the first to apply this literature to the military context. At the same time, another literature had evolved in the defense manpower arena that models individual retention over a military career, recognizing that retention decisions are made under uncertainty and that military members are heterogeneous in their tastes for military service. This retention modeling approach, known as the Dynamic Retention Model, was first developed at the RAND Corporation by Gotz and McCall (1984) to analyze Air Force officer retention; it was further extended and applied to other services, officers and enlisted personnel, active and reserve service, and the U.S. Department of Defense (DoD) civil service in later work. Chapter Six provides an overview of the DRM, where it is used to analyze the 40-year versus the 30-year pay table.

These two strands of literature—on structuring compensation across ranks in hierarchical organizations and dynamic retention modeling—were brought together and built upon by defense manpower economists, who developed a theory of compensation and personnel policy in the military context (Asch and Warner, 1994, 2001). As part of this work, they developed a theoretical framework that permitted an understanding of several puzzling differences between the military's compensation system and the systems typically found in large private-sector firms. Among those puzzles, and particularly relevant to the SASC's question, was the economists' observation that unlike large private-sector organizations, the military pay table—and, more broadly, the compensation structure—lacks skewness. *Skewness* refers to the observation that the difference in pay between ranks increases with rank, so that the pay levels of the most-senior leaders are substantially higher than those of senior and mid-level managers, while the pay of mid-level managers is only modestly higher than the pay of the lowest-ranked employees. Specifically, Asch and Warner (1994, 2001) observed that a typical O-6 receives about three times the basic pay of an O-1. By contrast, for example, Baker, Gibbs, and Holmstrom (1994) reported that level-6 managers in the private firm they studied earned five times more than level-1 managers. Other studies reported even more skewness, whereby pay in the upper ranks is many multiples of pay in the lower ranks (Main, O'Reilly, and Wade, 1993; Abowd, 1990; Leonard, 1990; Ehrenberg and Bognanno, 1990; Bognanno, 2001). In contrast, the chairman of the Joint Chiefs of Staff (four-star, rank O-10) earns only about 60 percent more basic pay than a typical O-6 colonel. The flatness of the military pay table relative to the private sector and the factors affecting the interrank pay spreads in the military was a puzzle that the theory presented in Asch and Warner (1994, 2001) was intended to address.[1]

This chapter reviews the arguments for a skewed compensation and the factors affecting the degree of skewness. It discusses why the military pay table is less skewed or flatter than what is typically found in large private-sector firms. The review provides a theoretical context for considering the question posed by the SASC about the usefulness of the 40-year pay table versus the 30-year table, and more broadly the question of how to set compensation to sustain personnel requirements in the upper ranks of the military. Reverting to pre-2007 compensation policies for senior personnel would effectively flatten military compensation for service beyond YOS 26 and further reduce the skewness of the military pay system. While theory cannot say how much skewness is needed, it provides insights into the factors to consider in setting pay in the upper ranks relative to the lower ranks.

A Theory of Compensation and Personnel Policy: Overview

The theoretical model in Asch and Warner (1994, 2001) addresses the question of how the level and structure of military compensation should be set, given the military's goals and objectives for manpower and personnel. A theory of compensation and personnel policy should incorporate the unique institutional features of the military, the labor supply and performance of military personnel given uncertainty about the future, and how the labor supply and performance respond to alternative compensation and personnel policies. This section provides a brief overview of such a theory, starting with a discussion of some key assumptions, and the next section focuses on the implications for skewness that follow from the model.

Personnel in the model are assumed to be heterogeneous. They vary with respect to their ability to perform tasks in the military and their taste for military life. While the military screens entrants, it cannot perfectly measure their ability or taste. It can only discern from retention behavior over a career that those who stay have a stronger taste than those who do not.[2] The military is also assumed to monitor individuals' work efforts only imperfectly. That is, the military may have difficulty discerning whether poorer performance is due to lower effort or to external factors that affect observed performance.

Organizational Objectives and Relevant Institutional Features

In any organization, including the military, the personnel system must meet the following objectives to be considered effective. First, it must attract and retain personnel in sufficient numbers to meet its skill and experience requirements and with sufficient quality and ability

[1] The theory also addressed two other puzzles related to differences between military and civilian compensation systems: the cliff-vesting of the military pension at 20 years of service in an immediate annuity and the more extensive use of up-or-out rules in the military.

[2] We can use the structure of the DRM together with data on retention over a career to estimate the parameters of the taste distribution of entrants (see Chapter Six).

to perform the required tasks. Second, it must train personnel and encourage personnel to acquire the skills they need to perform their tasks. Some of these skills require formal training, while others are learned on the job. The system must induce people to acquire both types of training, as required. Third, the system must motivate personnel to work hard and effectively. Because effort is not perfectly observed, the compensation and personnel system must provide adequate incentives to work hard and to seek advancement.

Fourth, the system must sort personnel and, in particular, must induce higher-ability personnel to stay and seek advancement. That is, it must set compensation and personnel policy in such a way that avoids adverse selection (where the least able stay), *climbing* (seeking ranks for which one is unqualified), and *slumming* (the converse of climbing). Finally, the system must induce personnel to separate when it is in the best interest of the services for them to do so. While people should stay long enough for the organization to obtain a return on training, the military wants people to separate, given its hierarchical rank structure and lack of lateral entry, to sustain promotion opportunities for those in the lower ranks. And, at some point, older personnel become less productive, though this point may be less relevant in the military, where even the top-ranked personnel rarely separate past age 60.

The military's ability to meet these objectives is affected by several organizational features. As mentioned, there is virtually no lateral entry to the active force. Thus, the senior leaders of the future must be recruited at entry. A second feature is the military's triangular hierarchical rank structure, in which positions in the upper ranks are filled from those at the next-lowest level. In the most-junior ranks, advancement is based on qualifications, subject to minimum time-in-grade and time-in-service requirements, and promotion decisions are made at the unit level based on the judgment of the local unit commanders. Beyond this level, future promotions are nationally competitive and based on centralized promotion boards. Promotion rates drop sharply, and competition for promotion becomes increasingly fierce as people move up the ranks, resembling a competitive contest or tournament. Third, the military relies on explicit up-or-out rules or mandatory separation rules for personnel beyond the junior ranks who fail to be promoted to the next rank. Such rules are rare in large private-sector firms, according to Baker, Jenson, and Murphy (1988) and are limited to academia and professional partnerships, such as law firms.[3]

Two important and related implications follow in the theory from the lack of lateral entry and the triangular structure of the hierarchy. First, the productivity of those in the upper ranks has a spillover effect on the productivity of those in the lower ranks. Poor performance in the upper ranks has a larger effect than poor performance in the lower ranks. Because higher-ranking personnel control more of the organization's resources and make decisions having greater overall impact, span-of-control considerations magnify the importance of having the most-capable personnel in the upper slots and motivating work effort among those in those slots. Second, personnel contribute to military readiness not only in their current position but also as potential selectees for future, higher-ranked positions. Thus, they have a productivity value equal to their productivity in their current position plus a shadow value associated with their productivity in future positions. As a result, new entrants at the bottom of the organization in the military must have higher ability on average than those in organizations that can

[3] Other institutional features are also relevant for setting military compensation and personnel policy, including the need to fill a heterogeneous set of jobs, often dangerous, across the globe with a variety of working conditions. We exclude these features here because they are less relevant to the skewness of the pay table.

fill senior positions directly from the external market. Furthermore, the military must design personnel and compensation policies to "percolate" the most able to the top.[4]

Individual Decisionmaking

Compensation and personnel policy must also consider the decisionmaking process of individuals and how those decisions respond to changes in compensation. The focus of the theory is on individual labor supply decisions (e.g., to join the military and to stay in the military) and on effort supply decisions (e.g., how hard to work). The decisions to join the military and to stay are modeled as in the DRM framework,[5] while the effort supply decision is modeled following the principal-agent models found in the economics and management literature.[6]

The DRM is a model of the decision to join the military and, for those who join, a model of retention decisionmaking over the life cycle in a world with uncertainty and where members have heterogeneous preferences (tastes) for active and reserve service. The DRM models members as forward-looking individuals who take into account future opportunities—in terms of current and deferred compensation in the military and in the civilian world—when making current retention decisions. So, they compare present and future military compensation with present and future civilian compensation, taking account of their underlying preferences for the military career compared with a civilian career and future uncertainty regarding environmental disturbances that can affect their valuations of military and civilian life.

Individual retention decisions result from a complex interaction of many influences. One of those influences is military compensation, not just one's current compensation but, because individuals in the model are forward-looking, one's future compensation as well. Similarly, the model includes external opportunities and specifically civilian compensation, both current and future, as an influencer of retention decision. The model recognizes that people are heterogeneous, as discussed above, and explicitly accounts for future uncertainty or unanticipated factors that affect the value of being on active duty, being in the selected reserve, or being a civilian. Examples are a good assignment, a dangerous mission, a strong or weak civilian job market, an opportunity for promotion, and a change in marital or health status. In the DRM, individuals are not bound by today's choice but can reoptimize in each future period, depending on future conditions as they become known. The flexibility to change one's future retention decision has value, and that value affects current retention decisions.

In addition to retention decisions, members make effort supply decisions in the model. Several factors affect this effort supply. Promotion to a higher rank provides monetary, nonmonetary, and intrinsic rewards. To the extent that future promotions depend on current performance, a reward in the form of a future promotion should induce individuals to work harder or more effectively in their current ranks. Individuals will also work harder in their

---

[4] As Willis and Rosen (1979) discuss, a complicating factor is that ability is not one-dimensional. Ability traits important for success in the lower ranks (e.g., physical strength or capacity to follow orders) may not be the same as those required in upper ranks (e.g., analytical reasoning or leadership skills). Skills that make one a good lieutenant may not make one a good colonel. If this is the case, performance in the lower ranks may not be a good predictor of one's performance in the upper ranks, making selection for promotion that much more difficult. The model does not treat this issue and assumes that ability is one-dimensional.

[5] The DRM has been developed and described in detail in a number of previous publications and will be discussed more in Chapter Six.

[6] The principal-agent literature is large. A representative review of this literature is Prendergast (1999).

current ranks the more they value the status and nonmonetary rewards associated with promotion. Importantly, monetary rewards associated with promotion can come in the form of pay (such as basic pay), avoidance of involuntary separation or up-or-out rules, or retirement benefits. Individuals may also work harder in their current rank if there is an intragrade payoff that is contingent on effort, such as a better assignment or a performance bonus. Intragrade performance incentives are weakened by the lockstep nature of the longevity pay increases in the military pay table. Finally, if promotion to a higher rank provides a signal to the external market about the individual's ability, effort today can increase the future value of the individual's expected alternative so that a reward to promotion may occur external to the military.

Skewness and the Sequencing of Intergrade Compensation Differences

The theory provides several insights about how pay should be sequenced across grades; in particular, it provides a rationale for a skewed sequence in which interrank spreads in compensation rise with rank.[7] We first discuss the rationale, and then the factors that tend to moderate the degree of skewness, thereby flattening the compensation structure across grades.

Rationale for Skewed Intergrade Compensation Differences

Over the initial years of service in the junior ranks, promotion rates are high and are generally based on members acquiring skills and meeting time-in-service and time-in-grade requirements. The theory implies that, in this case, intergrade pay increases do not need to be large to motivate effort. Indeed, we observe relatively modest average pay increases (given average time in service at promotion) in the early grades (Asch and Warner, 1996, Figure 10.2). However, beyond the early career, pay spreads begin to increase. Personnel begin to reach the middle ranks when promotions begin to resemble a tournament or contest for advancement. The military's objective is to sharpen the competition and induce the most qualified to reveal themselves in this contest. Larger pay spreads between ranks motivate harder work, discourage slumming, and encourage more capable personnel to remain in service and help maintain the quality of the promotion pool. Also, by improving the talent pool and by inducing the more capable to work harder, larger spreads prevent climbing.

As mentioned, at early- and mid-career ranks, promotion rates are high—that is, a high percentage of those eligible for promotion are eventually promoted. As individuals progress toward the senior ranks, promotion rates fall. Holding constant the size of the intergrade pay spread, declining promotion rates tend to decrease the expected payoff to advancement and

---

[7] These intergrade differences depend on the pay increases within a grade and, importantly, the timing of promotion, in addition to the sequence of differences in pay across grades given average promotion timing. However, for brevity's sake, we focus here on the intergrade pay spreads, assuming average times for promotion at each grade. Elsewhere (Asch and Warner, 1994, 2001), the roles of intragrade pay spreads and promotion timing in the structure of compensation are investigated. Also, as mentioned, incentives for effort and the sorting of higher-ability personnel can be influenced not just by pay but also by nonmonetary factors associated with promotion, plus elements of compensation that increase with promotion—notably, retirement benefits. The retirement system adds a degree of skewness to the compensation system, given the 20-year vesting in an immediate annuity, and permits a flatter pay structure in the pay table to produce the same retention and effort outcomes that would occur with a less generous system. Other elements of military compensation that could increase skewness are the basic allowance for housing and the tax advantage, both of which increase with grade. While we use the term *pay* in the following discussion of skewed pay structures, the discussion extends to include the other elements of monetary and nonmonetary compensation that increase with promotion.

thereby discourage effort and the sorting of higher-ability personnel. Interrank pay spreads need to rise with rank—they need to be skewed—to maintain effort and incentives for ability sorting.

The tendency to reduce effort is accentuated by other factors. The first is the rising importance of unpredictable factors in promotions to the higher ranks or factors that may be out of the individual's control. Performance in higher-ranked positions can be difficult to monitor and assess because of the skills, knowledge, and judgment needed for many decisions; performance is less related to explicit criteria and standards. In addition, promotion can depend on good fortune, such as the right assignments being available, helpful and well-regarded mentors, and external factors that make a decision turn out to be a success. The role of luck and the difficulty of assessing performance weaken the relationship between effort and the likelihood of promotion and, all else equal, discourage effort as individuals progress through the ranks.

The second factor that tends to reduce effort is the increasing homogeneity of the talent pool at higher ranks. In the lower ranks, there is likely to be a lot of variation in skill and talent among those available for promotion. When the promotion pool is heterogeneous, it is easier to bypass others through working harder. As individuals progress through the ranks, the pool becomes more homogeneous because of previous selections. Bypassing one's competitors by working harder becomes increasingly difficult, thereby blunting the relationship between effort and the probability of promotion and therefore blunting the incentives to supply effort. Another factor reducing effort incentives at higher ranks is the declining number of remaining promotion contests. As personnel progress through the ranks, the number of remaining promotions that can be earned, and therefore the number of promotion rewards, falls.

These arguments provide a rationale for skewed military compensation in which interrank spreads rise with rank. Skewed compensation offsets these factors that tend to discourage effort as personnel move to the upper ranks. Larger increases in compensation are needed in the upper ranks to offset the difficulty of monitoring and assessing performance, the growing role of luck in the promotion process, greater homogeneity in the talent pool, and the declining number of remaining promotions and subsequent rewards at higher ranks.

Factors That Decrease the Required Skewness

While skewed compensation can encourage effort, several factors decrease the skewness required. The more value that individuals attach to the status of the rank or to the nonpecuniary aspects of serving at a high rank, the smaller the monetary awards needed to motivate effort in the lower ranks. A second factor is the transferability of military experience and training. The less that training received in the military is transferable and improves outside employment opportunities, the smaller the in-service pay increases needed over a career to maintain a given level of retention.[8] The third factor is the role of taste for service and its correlation with ability. High-taste individuals are more likely to stay for future periods. Thus, hard work today has a higher expected future payout, for a given pay spread, because these individuals have a higher chance of being in the military to reap that payout. Therefore, higher-taste individuals will work harder in the current period. This implies that the higher the tastes of members,

[8] When human capital is general, so that training and experience are fully transferable, the alternative earnings stream in the external market is independent of the member's leaving date. But, in the absence of perfect transferability to other employers—when human capital is specific—the alternative earnings stream drops with length of service in the military. That is, those who stay longer would likely earn less in a nonmilitary career. Thus, sustaining retention in the presence of specific human capital requires less pay growth over the career.

the smaller the amount of skewness needed to induce a given level of effort. Furthermore, if tastes for service and ability are positively correlated so that personnel who have stronger tastes are also more capable, then the less skewness is needed to induce the more-capable personnel to stay and seek higher ranks. The personal discount rate also plays a role.[9] Those with higher personal discount rates, such as has been estimated for enlisted personnel (Asch, Hosek, and Mattock, 2014), will discount future rewards to promotion more, so that a given reward will have less impact on current effort. Put differently, we would expect officers to be more responsive to higher pay in the upper ranks than enlisted personnel, because estimates of officers' personal discount rates show that they are lower than those for enlisted. Thus, all else the same, the need for skewness is less for officers than for enlisted personnel. Finally, up-or-out rules can also induce effort by lowering the expected payoff to remaining in a lower grade. By inducing effort, these rules can serve as a substitute for increases in intergrade pay spreads. To the extent that such rules are stringent, especially in the more-senior ranks, the required skewness is less to sustain effort and induce ability sorting.

Thus, offsetting the factors that argue for more skewness are factors that include greater nontransferability of military experience to the civilian sector, more value placed on the nonmonetary aspects of promotion to the upper ranks, stronger tastes for service and a positive correlation of taste and ability, lower personal discount rates, and stringent up-or-out rules.

Insights for Comparing the 40-Year and 30-Year Pay Tables

The compensation and personnel policy theory provides a rationale for understanding why military compensation should be skewed as years of service increase and personnel promote to higher grades. Before 2007, no basic pay increases occurred after YOS 26 and no retirement benefit increase occurred after YOS 30. This pay table proved satisfactory in providing the services with a retention profile by years of service that met manning requirements, sufficient pools of qualified personnel from which to select senior leaders, and some number of personnel willing to continue beyond 30 years of service, depending on service needs.

However, several factors increased the demand for personnel to serve more than 30 years. These were the heightened manning requirements driven by military operations in the Middle East and in Afghanistan; increasing returns to technical knowledge and experience derived from increasingly sophisticated weapon systems, intelligence-gathering apparatus, and logistics systems, among others; and the realization that the military could gain from having GOFOs serve several assignments, rather than one, before leaving the military. The next chapter provides further information about these factors based on expert interviews, and Chapter Five reports tabulations of the number of officers, warrant officers, and enlisted personnel with more than 30 years of service, before and after the 40-year pay table was introduced in 2007.

To the extent that the demand for personnel with more than 30 years of service has increased and is expected to remain at its new, higher level, the implication of the theory is clear: Military compensation skewness should continue through the full range of service to

[9] The *personal discount rate* is the rate at which an individual values a dollar available one year in the future relative to a dollar available today. For example, a personal discount rate of 0.10 would imply that a dollar next year is worth $0.909 (1 / [1 + 0.10]) today. A high personal discount rate implies that future compensation has a much lower value to the individual than current compensation.

40 years, not end at 30 years. Foremost, this will sustain the incentive to remain in the military. It will also sustain—not decrease—the incremental incentive to exert effort and reveal ability. But as we have noted, the most-experienced military personnel, those serving beyond 20 years, are already highly selected; the skewed compensation system has, according to theory, induced selection on taste, effort, and ability. Therefore, as one looks from 20 to 30 years of service forward to 30 to 40 years of service, these factors imply that skewness could be lower than if there was no such selection. But theory does not say how much less.

Finally, despite skewness being a central concept of the compensation and personnel policy theory, it is important to recognize that the 40-year pay table, by itself, delivered little skewness, especially for senior officers. As the figures in Chapter Two showed, basic pay increases were primarily for E-8s and E-9s, W-5s, and O-8s; increases for O-9s and O-10s, although formally present in the pay table, were largely unrealized because of the Executive Level II pay cap. Instead, skewness for senior officers came from the legislative provisions dealing with retirement benefits. Here, the major source of increase came from making years of service past 30 countable in computing the retirement benefit (retirement benefits are 75 percent of basic pay at 30 years and 100 percent at 40 years). Basic pay increases within grade, as well as promotion to higher grades, also increased the retirement benefit.

This discussion leads to the question of how much skewness is enough—that is, how flat or skewed should compensation be? The theory provides broad guidance on the factors that lead to a more or less skewed military compensation system, though many of these factors are not easily quantifiable. It seems unlikely that the optimal structure would mean no skewness in the upper ranks among those most-experienced personnel. That said, ultimately, assessment of the adequacy of the compensation system, including its degree of skewness, must rely on assessments of whether the system meets the organization's objectives of attracting, retaining, sorting, motivating, and eventually separating personnel. We do not know of past studies that provide such quantitative assessments, especially as they pertain to upper-ranked military leadership positions. Our interviews and analysis using the DRM aim to provide information about retention experience before and after the change to the 40-year pay table and about what would happen if the pay table reverted to 30 years (in terms of stopping longevity increases beyond YOS 26) and the retirement-related provisions were, or were not, also eliminated.

CHAPTER FOUR

Results: Major Themes Emerging from Interviews

Approach

We conducted 21 interviews with senior military officers and civilian DoD employees in positions of authority relevant to military compensation, retention, and personnel and force management. These experts were from a cross-section of the services and from various offices within OSD. In selecting interviewees, we considered several criteria. First, we hoped to interview representatives from each service, the Joint Staff, and the reserve components. We focused on those military personnel—officers and senior enlisted—who had experience and involvement in personnel issues. We also wanted to include civilian personnel working on personnel issues within OSD. We tried to include officials involved in as many different aspects of personnel management as possible, both those responsible for these issues now and those who were responsible for these issues at the time of the 2007 pay table change. Our sponsor also contributed to defining our interview sample and offered guidance on who might provide the most-valuable and -relevant insights. The sponsor was also responsible for contacting and setting up the interviews. We used a semistructured interview protocol, allowing us to gather the perspectives of our interviewees on three broad topics:

1. overall assessment of the performance of the 40-year pay table versus the 30-year table; advantages of the 40-year versus 30-year pay table in meeting current and near-term retention goals
2. changes in force management as a result of the 40-year pay table
3. whether returning to the 30-year table would necessitate changes in policy, such as using incentive pays.

Thus, the focus was on the benefits of the longer pay table, the effects it has had on retention or personnel management decisions, and the possible effects on retention that would result from a change back to a 30-year pay table. The semistructured nature of our interviews allowed us to explore the unique experiences and observations of our interviewees, in both their current and past positions. As noted, the change in the pay table came with several other legislative changes that affected compensation, particularly for senior officers. Our interview questions focused specifically on the pay table change, but many interviewees also discussed the other changes and their effects on retention and personnel management. Few interviewees provided analysis or data to support their observations. Thus, their input was based on the sum of their experience, including past analyses they had seen, observation, and speculation.

While perspectives and observations differed among the experts, several themes emerged from the interviews. We summarize them here.

Intended Objective of the Change to the 40-Year Pay Table

Several of the interviewees provided background on the objectives of the change to the 40-year table. They shared the view that the longer pay table was intended to increase retention of senior officers and, to a lesser extent, senior enlisted personnel. Then–Secretary of Defense Rumsfeld was a strong advocate of the change and made comparisons to the private sector's approach to compensation. One interviewee commented, "Rumsfeld challenged his OSD leaders on why the military would encourage departure when people are at the peak of their professional contribution, unlike the private sector." Given the role of the pay table as a tool for talent management, the change to the longer pay table was intended to increase the ability of the services to retain personnel whose experience made them invaluable to the military. As one interviewee asked, echoing Rumsfeld's argument above, "Why should we get rid of someone just because of how long they have been in the service?"

Interviewees noted that it was important to view the pay table change in its historical context. It occurred during a period of high strain on the military and high demand for personnel. The wars in Afghanistan and Iraq were both ongoing, and the services, especially the Army and Marine Corps, were hard-pressed to bring in and keep enough personnel. Retired personnel were being called back, and the reserves and National Guard were being used heavily. In fact, one interviewee noted that recalled retirees, many of whom already had 30 or more years of service, were one of the target populations of the longer pay table. This interviewee explained, "The 40-year table provided an incentive for these recalled people." With the conflicts in Afghanistan and Iraq came more three- and four-star billets for jobs directly related to the ongoing conflicts, thus creating more demand for three- and four-star generals. The increased demand for senior officer billets was thus another motivation for policy changes aimed at retaining senior and experienced personnel.

Performance of the 40-Year Pay Table

Virtually all of the interviewees stated that senior officers serving more than 30 years do not remain in the military for the money and are not motivated by the small pay increases earned for longevity past 30 years. They argued that these personnel instead serve to support the mission, are intrinsically motivated and highly selected, and stay as long as they are asked to stay. That said, the interviewees also agreed that the pay table is one component of an overall compensation package and that it is critical to view the package in terms of how well it enabled the services to manage talent, not just those with more than 30 years of service but also more-junior personnel coming up the ranks who might fill those positions in the future. Within this context, the interviewees differed on the importance of the pay table in the overall package; whether the move to the 40-year table had an effect on retaining those with fewer, and more, than 30 years of service; and whether any changes in retention after the move to the 40-year table in 2007 could be attributed to the shift to the 40-year table.

Interviewees gave various reasons for why the move to the 40-year table had little effect, from their perspectives. Some interviewees gave greater weight to the argument of intrinsic motivation and highly selected personnel. For that reason, from their perspective, the change to the 40-year table had little retention effect as far as they could discern. Others stated that the 40-year table had no effect but that other compensation changes, particularly lifting the pay cap for calculating retirement pay, did affect the retention decisions of senior officers. Some interviewees argued that the pay table had few effects on the number and types of people serving past 30 years because the services had the tools needed to keep these personnel when necessary. Consequently, the services could have managed any retention issues that arose.

Some interviewees stated that force size increased after 2007 among those with more than 30 years of service but argued that those increases could be attributed to factors other than the change to the 40-year pay table. Factors they mentioned included the economic downturn that began at the end of 2008, the pace of deployments, the high demand for personnel during recent operations in Afghanistan and Iraq, and such policies as stop-loss and the recall of retirees to help meet that demand.

Other interviewees perceived that the move to the 40-year table did affect the number and types of people who stayed past 30 years of service. These interviewees argued that the longevity increases provided between 30 and 40 years of service affected the retention decision process of senior personnel, often involving the input of their spouses. Some cited anecdotes and spoke from personal experience. Many also stated that although the longevity increases were not large, they were a way of recognizing senior personnel and their families for their service, sacrifice, and commitment. This recognition affects morale and thus retention incentives. Several interviewees noted that chief executive officers in the private sector make millions of dollars, so senior officers, while they cannot be compensated at this level, deserve to be compensated fairly for their sacrifice.

Some interviewees commented that the change in the pay table gave the services greater flexibility to keep personnel longer than 30 years, even if the flexibility is not always used. They noted that the services still had the ability to get rid of personnel when necessary, because of time-in-grade limitations and review boards. Others commented on the effects of the 40-year table on personnel costs. Because the change in the pay table affected a small group of individuals, its overall cost was small relative to overall basic pay and personnel costs. Finally, some mentioned the effect of greater retention among senior personnel on the promotion pipeline for more-junior personnel. These interviewees' concern was that increasing retention could create a logjam in promotion potential, although none provided any information about whether this had occurred.

Necessity of Keeping the 40-Year Pay Table

Most interviewees said that Congress could get rid of the 40-year pay table, replacing it with the 30-year table, without having much of an impact on retention—assuming none of the other 2007 compensation changes were reversed and as long as the services maintained the tools needed (including special and incentive pays) to flexibly manage any impact that arose. Interviewees offered a variety reasons for why they thought getting rid of the table would have little effect, citing the factors listed above. At the same time, nearly all of the interviewees said that they would not recommend that Congress get rid of the 40-year table, again for a range of reasons.

The reasons cited for why changing back to a 30-year table would have little effect, assuming no other compensation changes occurred, are essentially the same reasons why interviewees thought the move to the 40-year table in 2007 had little effect. For example, many felt that senior officers stay in service because of their commitment to the mission and because of the opportunity for promotion to specific positions. One interviewee commented, "The 40-year pay table is not a motivator for general officers. Most are selected for Brigadier General at 24 or 25 years of service, and they leave when the Vice Chief tells them. Otherwise they stay. That is, they stay as long as there are current and future assignments available to them." Several interviewees also argued that even when highly skilled and experienced personnel retire, there are always highly qualified junior personnel ready to step up and take their place. As one interviewee commented, "Cases where someone is uniquely qualified are pretty rare."

Similarly, interviewees were willing to accept a change back to the 30-year pay table as long as the services had the flexibility to use incentive pays, such as assignment incentive pay, to retain personnel as needed. They wanted DoD to take a more careful look at DoD instructions that limit the use of incentive pays to provide the services with more flexibility.

Reasons Cited for Not Reverting to the 30-Year Table

One reason frequently given by interviewees for staying the course and not reverting to the 30-year table was the perception that doing so would have a negative effect on morale and the retention decisionmaking process, especially among spouses. Some interviewees felt that the effect of the change would be most strongly resented by senior officers, coming on top of several pay freezes and a reinstatement of the pay cap in 2014 for the purpose of retirement calculations. They saw that such a pay table change would be seen as "adding insult to injury" and as a sign of disrespect for the country's senior military leaders and the sacrifices they had made for the country. One interviewee asked, "How many times can you kick senior officers in the gut?" For many interviewees, the longevity pay increases past YOS 30 serve as symbols of appreciation for the sacrifice of senior personnel. An interviewee commented, "We need the bumps [longevity increases] in the 40-year pay table as a recognition of service even though amount is nominal."

A related reason offered was that reinstating the 30-year table, though having a small perceived effect on retention, would have a large negative effect on service member perceptions about the stability of their pay and benefits. Given the recommendations by the Military Compensation and Retirement Modernization Commission regarding changes to commissary benefits, basic allowance for housing, and health care, as well as the recent reform to the military retirement system, a change back to a 30-year pay table could become part of a larger narrative about how Congress is cutting military pay and benefits. Such perceptions, many interviewees felt, can hurt morale, and therefore retention and recruiting.

Yet another reason that some interviewees offered was the lure of a civilian career for senior officers. They noted that at age 50 or 55, it is possible for service members to leave the military and pursue a civilian executive job. However, these opportunities dwindle as age advances past 55. Interviewees felt that without longevity raises and a sense of appreciation of the value their service, members might choose to leave earlier rather than later. Other interviewees felt that even for those uninterested in a civilian career, the loss of longevity raises, combined with changes to the retirement calculation, would cause individuals to view service past 30 years as "not worth the additional sacrifice required of senior personnel and their families." Still others argued that the retention effect could trickle down to junior officers who might feel that the changes in senior officer compensation signaled that a long-term commitment to the military was not valued or rewarded.

Some interviewees commented that the effect of returning to a 30-year pay table would likely be small for officers but could be significant for senior enlisted and warrant officers.

These personnel earn less money than officers, and the lack of longevity increases beyond YOS 30 could have more of an effect on their earnings and their retention. Some interviewees were concerned that even if a shorter pay table did not affect the number of people who stayed past 30 years, it might have an effect on the quality of personnel. One interviewee noted, "You might not end up with the best people in the open jobs."

Interviewees who thought that keeping the 40-year table was a good idea often cited the importance of having the flexibility to adjust compensation for very senior members, even if there has been little necessity for that flexibility in recent experience. They emphasized the lack of lateral entry to the senior ranks and the difficulty of replacing the experience and knowledge embodied in personnel at that level via other means. Several noted that senior enlisted and warrant officers with more than 30 years of service often had irreplaceable technical expertise, while senior military officers (e.g., the Chairman of the Joint Chiefs, combatant commanders) had leadership skills, knowledge, and experience that would be difficult to replace. One interviewee said, "If pay is capped at 30 years, we'll have to scramble to get younger, less experienced people into these billets, and that tends to hollow things out." The 40-year pay table was described as a "good tool to have in the toolbox," and as "insurance for keeping people." Some noted that a 30-year table could also be used, but it would be important to have flexible tools to retain key personnel when necessary. Some interviewees stated that the shift to a 40-year pay table did allow three- and four-star officers to have multiple assignments, as Secretary Rumsfeld argued it would, and the flexibility to accommodate multiple assignments in the future was an important reason for keeping the 40-year table.

More broadly, some interviewees stated that longer careers among senior personnel improved military readiness and made good financial sense. The military makes significant investments in training and developing these personnel, and longer careers allow a longer return on the investment. These interviewees noted that senior personnel are the ones who train junior personnel, so a loss of senior personnel would be a double blow to readiness: the loss of the personnel themselves and the loss of quality trainers to prepare future personnel.

Related to readiness and flexibility, others pointed to the changing nature of military service as an argument for keeping the 40-year pay table. Careers are getting longer as people live longer, and military service is becoming less physically oriented and more technically oriented. This may make it possible for members to have longer productive careers and may also raise personnel quality and technical skills in such fields as cybersecurity and cyberwarfare. One interviewee noted, "The types of people that we need to execute a war are completely different than in the past. We need technical experts. This means we are likely to make more use of senior enlisted and senior officers who have this expertise."

Some interviewees supported keeping the 40-year table because reverting to a 30-year table would have potential negative effects on O-6s with prior enlisted experience. Under the 30-year table, most such O-6s retired at YOS 26 or 27, maybe staying until 30 years of service to get the maximum retirement benefit. There was no real incentive to stay past 30 years, and, in any case, time-in-grade limitations required them to retire at 30 years of service unless they received a waiver, which was rare. In contrast, under the 40-year pay table, O-6s with prior enlisted service had a stronger incentive to stay past 30 years, and given the wars in Afghanistan and Iraq, the services allowed more of them to do so. Returning to a 30-year pay table could hurt the retention of these personnel, some interviewees stated. Yet one interviewee noted that the motivation for changing to the 40-year pay table was not to lengthen the careers of O-6s with prior enlisted service.

Finally, several interviewees noted that going back to the 30-year table would be logistically and administratively cumbersome, would possibly require policy revision or new legislation, and would hurt members' perceptions of a predictable, steady element of pay that could factor into retention decisions in a systematic way. A number of interviewees stated that the additional cost of the 40-year table was so small relative to the overall compensation budget that it was not worth creating negative perceptions among members about their pay and benefits.

Additional Issues

A few interviewees touched on other issues. The first is the potential for differential effects on the active versus the reserve component of a 40-year versus 30-year pay table. Some interviewees noted that because reservists can have 25 to 30 years of service and not yet be 60 years old, more people may get to 40 years of service by the time they are 60 in the reserve component than in the active component. Still, the effect of a change might not be all that great even for these individuals, because in the reserves, most officers are capped at a two-star level.

A few interviewees asked whether additional retention and experience were necessarily desirable, given the budgetary cost and the potential for slowing the promotion pipeline for more-junior personnel. Furthermore, some interviewees were not sure whether additional retention would have a uniformly positive effect on military readiness. One interviewee asked, "Is an O-6 with 32 years of service really that much better than an O-6 with 26 years of service?" Others raised the issue of officer management. Several interviewees felt that considerations about changes to the pay table should be considered alongside retirement reform and Defense Officer Personnel Management Act (DOPMA) reform. One interviewee commented, "I wouldn't touch anything until retirement reform is done," and adding later, "Don't touch the pay table until a DOPMA review."

Cap on Basic Pay for Computing Retired Pay

Finally, many interviewees had strong views about the 2015 NDAA provision that reinstated the executive pay cap on the retirement benefit calculation, effective for service after December 2014. Nearly all of these interviewees stated that they believed this change would affect the retention decisions of senior officers. One interviewee noted, "It's about the caps. . . . The pay table itself is not an issue either way." Several interviewees noted that they have already heard of senior and junior officers considering separating from the military because the change in the retirement calculation reduced the incentive to stay. One commented, "With the cap on retired pay, people are more likely to leave earlier and pursue a corporate job than stay longer [in the military] and pursue a nonprofit job." An interviewee said that while senior officers may choose to stay given the amount of time they have already invested, the loss of junior officers is a stronger possibility and a more serious problem, noting, "We will lose good people."

Many of these interviewees were adamant that this legislative change should be reversed if the services are to sustain retention of their senior officers. Others were more cautious, arguing that the need to retain senior officers depended on overall requirements for such officers and how important they were to readiness, whether the retention of up-and-coming officers was sufficient to meet those requirements, whether the more-junior officers had the required expertise, and, finally, how much flexibility is needed today and in the future to access the type of expertise embodied in these most-senior officers.

Summary

Overall, most interviewees did not think that the change to the longer pay table in 2007 had significant implications for retention or personnel management, and most did not think that going back to the 30-year table would pose significant retention challenges in the near future. Yet nearly all thought that going back to the 30-year table was a bad idea, offering a diverse array of explanations for this view. Furthermore, while nearly all interviewees believed that the change to the 30-year table was a minor issue, nearly all considered the reinstatement of the cap on pay for the purpose of computing retired pay to be a major issue. While most of the interviewees supported reversing the 2014 change, some stated that more information was needed before they could support such a reversal.

CHAPTER FIVE

Results: Trends in the Retention and Strength of Senior Military Personnel

The SASC asked that DoD's assessment of the pay table contain "a description of how many personnel remained on Active Duty past 30 years of service, annually since 2007; the breakdown by pay grade of such personnel; and the additional costs to the Department of Defense since 2007 of operating under the 40-year pay table rather than a 30-year pay table" (SASC, 2014). This chapter addresses the first part of the SASC request, by discussing trends in the number of military personnel, by grade and service, staying in the military past 30 years of service from 2000 to 2014. (Chapter Six discusses costs of operating under a 40-year versus 30-year table.) The discussion here considers not only these trends but also the historical context and factors other than the pay table change that may have affected the number of personnel with more than 30 years of service. It is important to note that while the trends illustrated here provide information on the number of people affected by the policy change and how that changed over time, they do not provide a causal analysis of how policy changes affect retention.

Data and Approach

We used the Active Duty Pay file from the Defense Manpower Data Center for our tabulations—specifically the September inventory of personnel from 2000 through 2014, including only regular component personnel (and not activated reservists). We created cross-tabulations by year of service; by pay grade and year of service; and by pay grade, service, and year of service. Years of service for the purpose of our tabulations are computed using pay entry base date. We also looked at continuation rates, or the rates at which personnel choose to stay in the military at each year of service greater than 20.

We encountered two data anomalies in the Active Duty Pay file worth noting at the outset. First, there were a handful of individuals in low-ranking grades, such as E-3, E-4, O-2, and O-3, who appear in the data to have more than 30 years of service. Because it is unlikely that an individual would still be an E-3 or E-4 after serving in the military for more than 20 or 30 years, and in most cases impossible due to time-in-grade limitations, these outliers are likely data-coding errors. However, the number of these personnel in the data is fairly small and unlikely to significantly affect the overall trends in the number of personnel in senior officer and enlisted ranks with more than 30 years of service since 2007.[1] Second, Marine Corps data

[1] We assume any other coding errors to be randomly distributed through the data and not likely to affect our overall assessment of trends.

in the Active Duty Pay file for the years 2000 and 2001 did not distinguish between regular and reserve-component personnel and are therefore missing from our analysis. However, based on the number of Marines with more than 30 years of service in the other years in our data, the number of personnel omitted is likely only a small percentage of the total and unlikely to affect overall trends.

Results

Interpreting Trends in Historical Context

When interpreting trends in the number of personnel with more than 30 years of service, it is important to consider not only the pay table change in 2007 but also the many other factors that may have affected the number, length of service, and grade mix of military personnel over the period under consideration. First, there are the other legislative changes that also occurred in 2007, as discussed in Chapter Two. These included the change in the cap on senior officer pay to Executive Level II and the removal of the pay cap in retirement calculations, as well as the pay increases afforded by the longer pay table. As shown in Chapter Two, the increases in pay afforded by the longer pay table were actually fairly small, except for W-5 personnel. The financial implications of the changes to retirement calculations were more significant. This increase in retirement benefits exerted a pull and a push on personnel deciding whether to stay or leave the military. The retirement benefit for those staying beyond 30 years increased especially, because years after 30 became countable toward retirement, which increased the incentive to stay. However, the lure of these higher retirement benefits also increased the incentive to leave the military to begin claiming them.

Second, 2007 was a period of high demand on military forces for the wars in Afghanistan and Iraq. The military needed additional personnel and used emergency authorities to get them, including recalling retired reservists (who often already had 30 years of service) and invoking stop-loss, which kept personnel in the military past the end date of their expected commitment. The military also used waivers that allowed personnel to serve past time-in-grade limits (e.g., past 30 years for O-6s) and that permitted it to temporarily exceed caps on the number of flag officers and senior enlisted personnel. The ongoing wars in Afghanistan and Iraq may also have affected incentives to stay among military personnel. Some may have had an interest in extending their service to contribute to the mission or to see the conflicts out to their ends. Furthermore, as noted in the interviews, the wars in Afghanistan and Iraq increased the demand for senior officers by creating new billets associated with the ongoing campaigns.

Finally, the state of the economy may also have affected the number of personnel choosing to stay in the military. The great recession began in 2008 and severely reduced economic opportunities in the civilian labor market and made a continuing career in the military more appealing for senior personnel who, as they reached 30 years of service, may have been considering a civilian job. This may also have affected the number of personnel in the military with more than 30 years of service.

Summary Tabulations

Overall, our tabulations show that the number of personnel with more than 30 years of service has increased since FY 2007. As noted, this increase cannot be attributed exclusively to the pay table change, given the other factors operating at the same time. Also, the overall increases in personnel mask more-subtle differences among the services and among officer, warrant officer, and enlisted personnel. These trends are also discussed below and in the appendix to this report.

Figure 5.1 shows the number of active-duty personnel with more than 30 years and more than 35 years of service from FY 2000 through FY 2014. The figure and data labels show that the number of personnel with more than 30 years of service has increased since FY 2007. There has also been a slight increase in the number of personnel with more than 35 years of service. In FY 2005, there were approximately 3,972 personnel with more than 30 years of service, and by 2014, there were 6,583, for an increase of 66 percent over that period. The upward trend in the overall number of personnel with more than 30 years of service appears to begin in 2006, accelerating in 2007 and continuing through the present.

The increase in the number of personnel with more than 35 years of service is smaller, though still considerable in relative terms. In FY 2005 and FY 2006, there were 615 and 608 personnel with more than 35 years of service, respectively. This peaked at 871 in FY 2013, representing a 42-percent increase between 2005 and 2013. The number declined by 8 percent between 2013 and 2014, from 871 to 799.

Figure 5.1
Number of Active-Duty Personnel with More Than 30 Years of Service

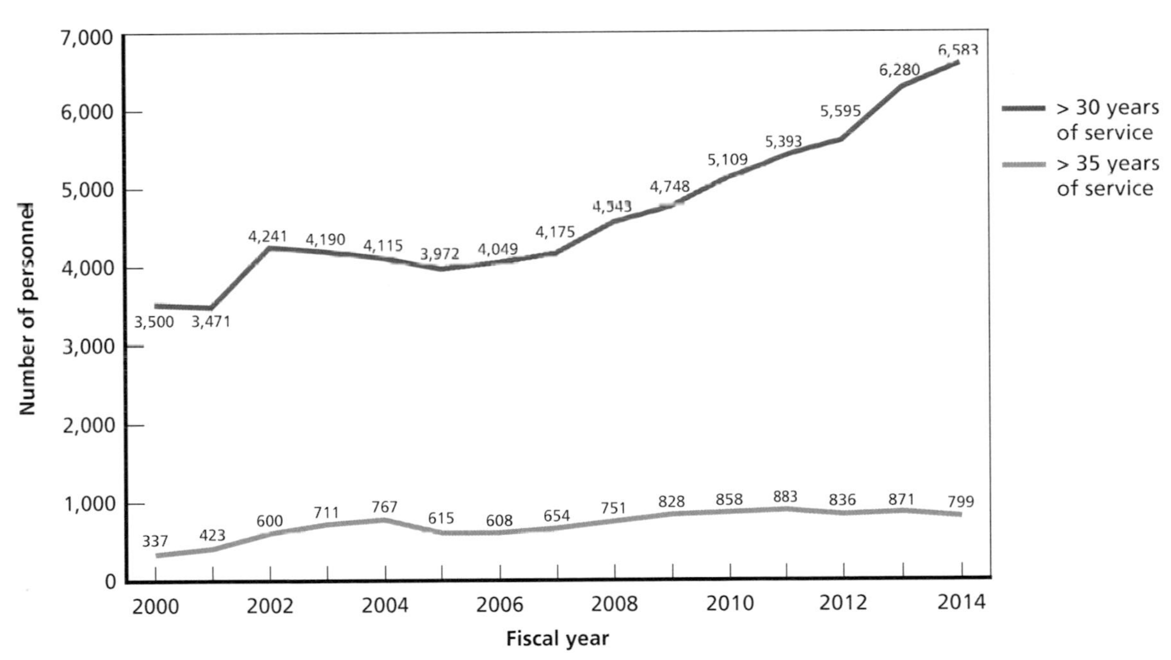

NOTE: Totals for 2000 and 2001 do not include Marine Corps data, because data in these years did not distinguish between active and reserve personnel. This is true of all force-wide trends, as well as the trend graphs focused on the Marine Corps.

RAND *RR1209-5.1*

Tabulations by Pay Grade

We also consider trends in the number of active-duty personnel with more than 30 years of service for officers, warrant officers, and enlisted personnel separately.

Figure 5.2 shows the total number of officers with more than 30 years of service, while Figure 5.3 and 5.4 show the number of O-4 to O-6 personnel and O-7 to O-10 personnel, respectively, with more than 30 years of service. Any O-6 or below with more than 30 years of *commissioned* service must receive a waiver to continue serving, because of time-in-grade limitations. Many O-5s and O-6s with more than 30 years of service may be personnel with prior enlisted experience who do not yet have 30 years of commissioned service.

Figure 5.2 shows that the number of officers with more than 30 years of service has increased since 2007. In FY 2005, there were 2,901 officers serving past YOS 30. This number fell slightly in FY 2006 to 2,861 but by FY 2014 had reached 4,019, for a 41-percent increase between 2007 and 2014.

Most of the observed overall increase in Figure 5.2 comes from the increase in officers in grades O-4 through O-6 (Figure 5.3), and not in the more-senior ranks (Figure 5.4), which was the goal of moving to a 40-year pay table. This increase in O-4s through O-6s may be driven primarily by individuals with prior enlisted service who faced stronger incentives to stay following the pay table and legislative changes and perhaps greater ability to stay as the demand for personnel rose with the conflicts in Afghanistan and Iraq. In this case, the pay increases that came after 30 years of service amounted to a "bonus" for officers with prior enlisted service.

Looking at Figure 5.3, the number of O-6s with more than 30 years of service stood at 1,460 in FY 2005 and peaked at 1,611 in FY 2013. In FY 2014, there were 1,580 O-6s with more than 30 years of service. The increase has been more dramatic for O-5s. While there

Figure 5.2
Number of Officers with More Than 30 Years of Service

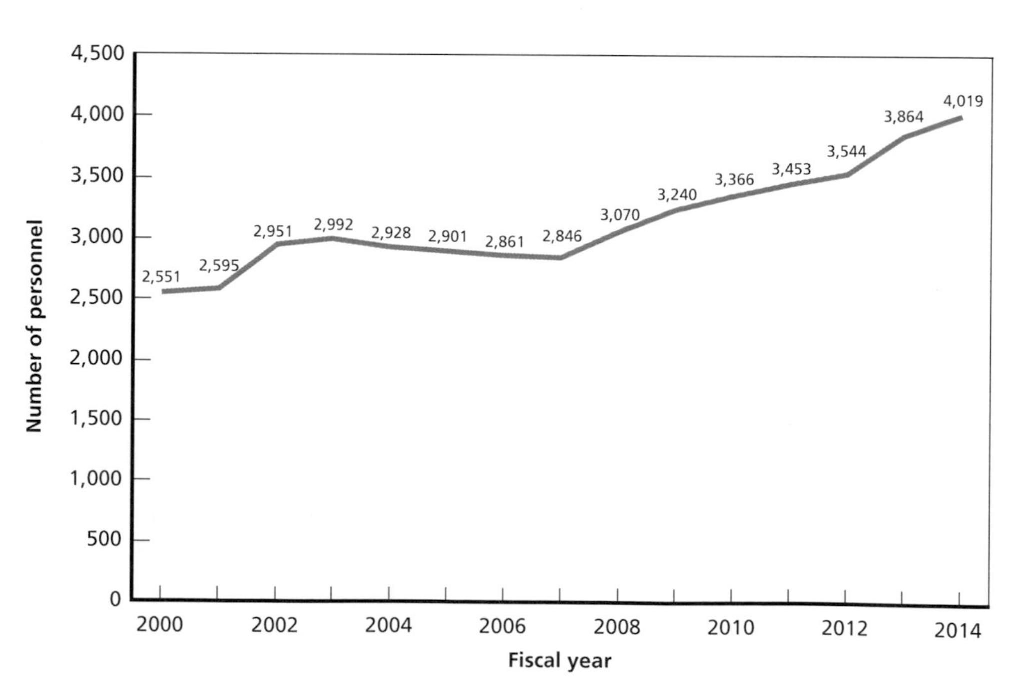

RAND *RR1209-5.2*

Figure 5.3
Number of Officers with More Than 30 Years of Service, O-4 to O-6

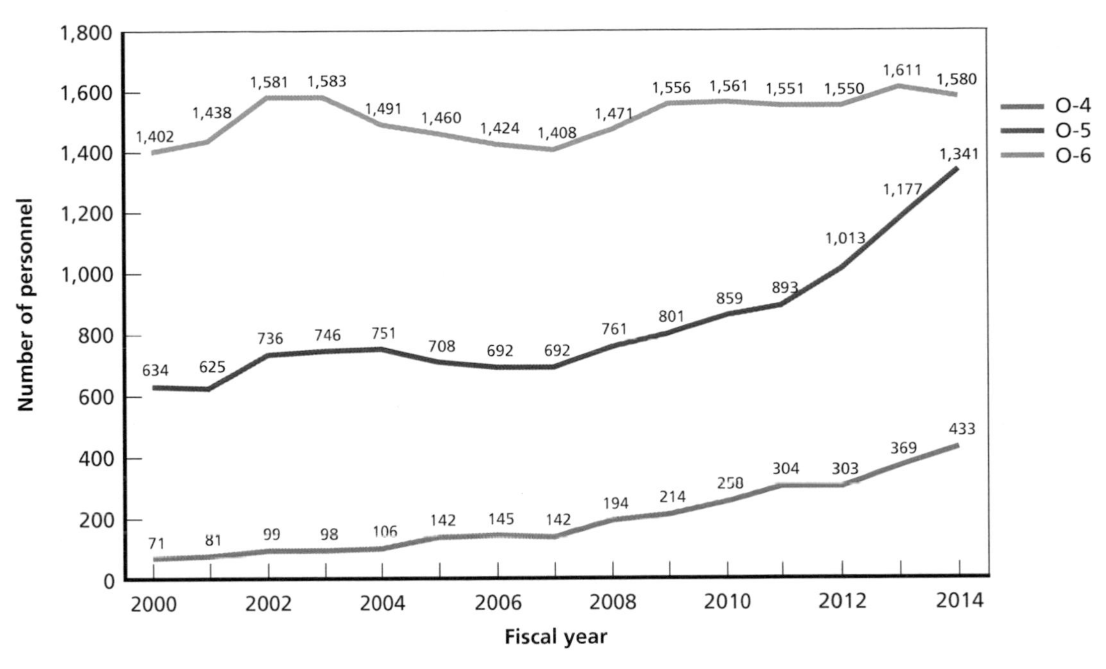

RAND *RR1209-5.3*

Figure 5.4
Number of Officers with More Than 30 Years of Service, O-7 to O-10

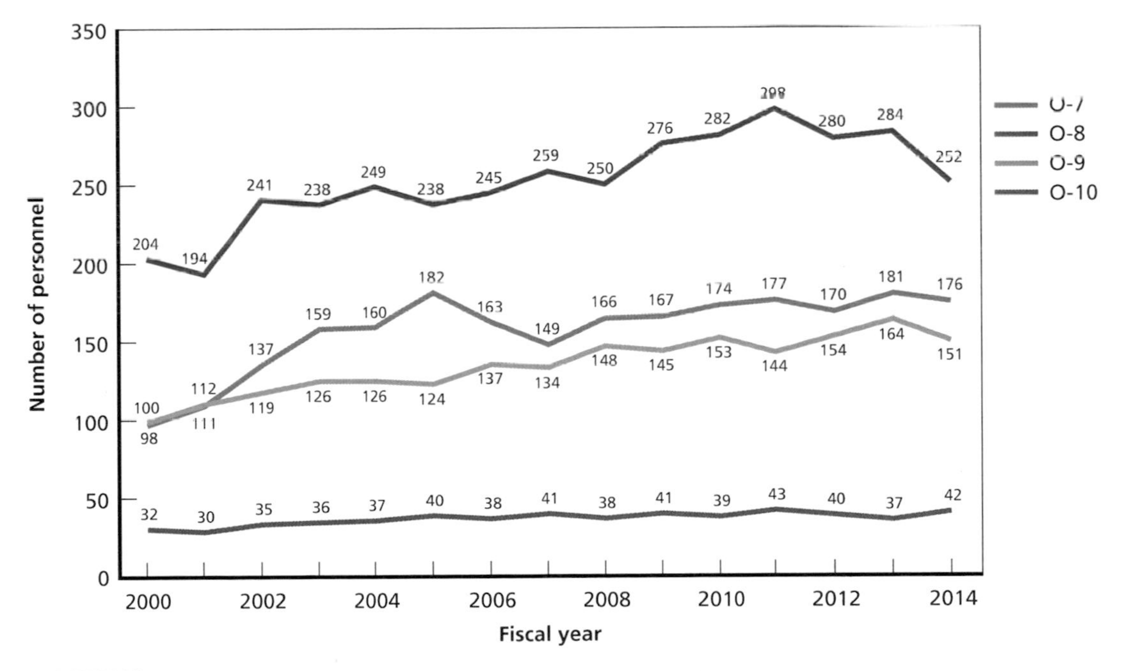

RAND *RR1209-5.4*

were 692 O-5s with more than 30 years of service in FY 2006, this number rose to 1,341 in FY 2014. The number of O-4s with more than 30 years of service also increased, from 145 in FY 2006 to 433 in FY 2014. Overall, the number of active-duty personnel in O-4 to O-6 increased from 2,242 in 2007 to 3,354 in 2014, a 50-percent increase.

In contrast, Figure 5.4 shows that the number of senior officers who have more than 30 years of service increased far less. Between FY 2007 and FY 2013, the number of personnel in O-7 to O-10 increased from 583 to 666, or 14.2 percent, while between 2013 and 2014, the number fell to 621, a decline of 6.8 percent. Overall, between 2007 and 2014, the number of senior officers increased by 6.5 percent, far less than the 50-percent increase in the number of O-4 to O-6 personnel.

For O-10s, there has been essentially no change between FY 2007 and FY 2014. For O-9s, the number has risen from 124 in FY 2005 to 164 in FY 2013, then fallen back to 151 in 2014. The number of O-8s had been increasing gradually since about 2001. It stood at 245 in FY 2006 and rose to 298 in FY 2011, then fell to 252 by FY 2014. Finally, the number of O-7s with more than 30 years of service peaked at 182 in FY 2005 and fell to 176 by FY 2014.

Figures 5.5 and 5.6 offer details on the number of officers with more than 20 years of service. Although the number of officers with more than 20 years of service has increased since 2007, this continues an upward trend that began in 2001. Furthermore, the increase appears to have leveled out and decreased slightly since about 2010. The majority of this increase has come from the O-4 to O-6 grades and not O-7 to O-10 grades. In fact, the number of O-7s to O-10s increased only from 902 in FY 2004 to a peak of 1,003 in FY 2010 before falling to 945 in 2014.

Figure 5.7 shows the number of warrant officers with more than 30 years of service, and Figure 5.8 shows the number by pay grade, W-3 to W-5. The number of warrant officers staying past 30 years of service has increased since 2007. While there were 376 warrant officers with

Figure 5.5
Number of Officers with More Than 20 Years of Service, All Officers and O-4 to O-6

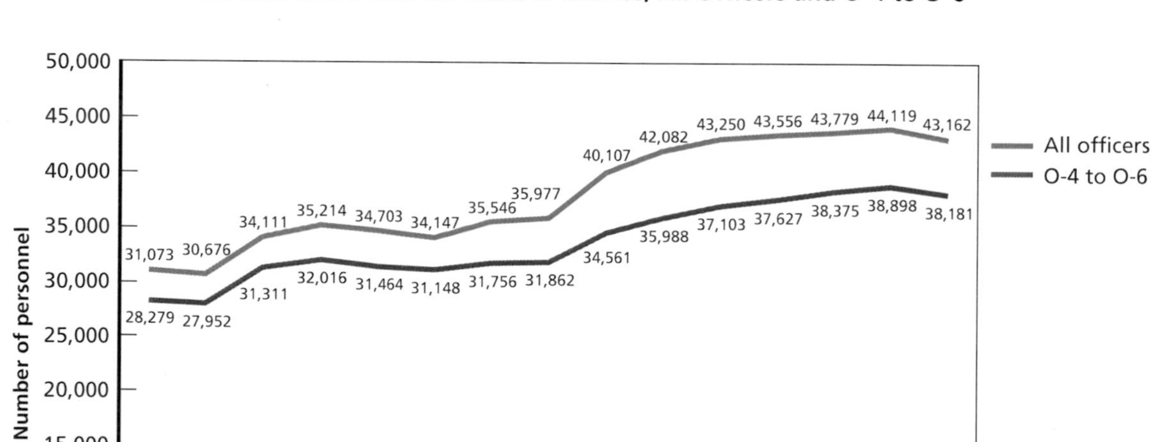

RAND *RR1209-5.5*

Figure 5.6
Number of Officers with More Than 20 Years of Service, O-7 to O-10

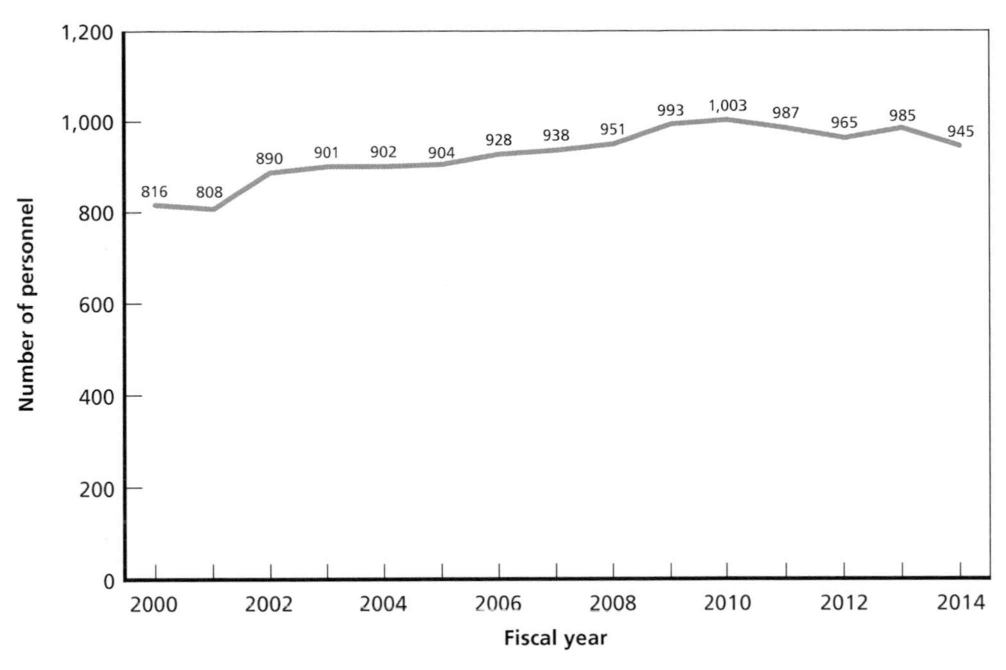

RAND *RR1209-5.6*

Figure 5.7
Number of Warrant Officers with More Than 30 Years of Service

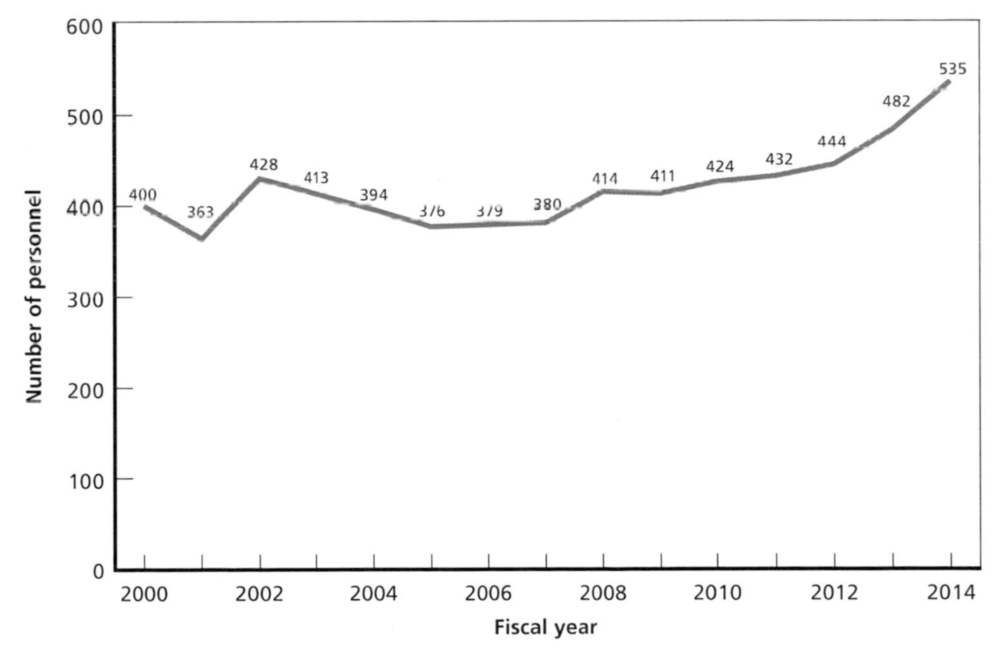

RAND *RR1209-5.7*

Figure 5.8
Number of Warrant Officers with More Than 30 Years of Service, W-3 to W-5

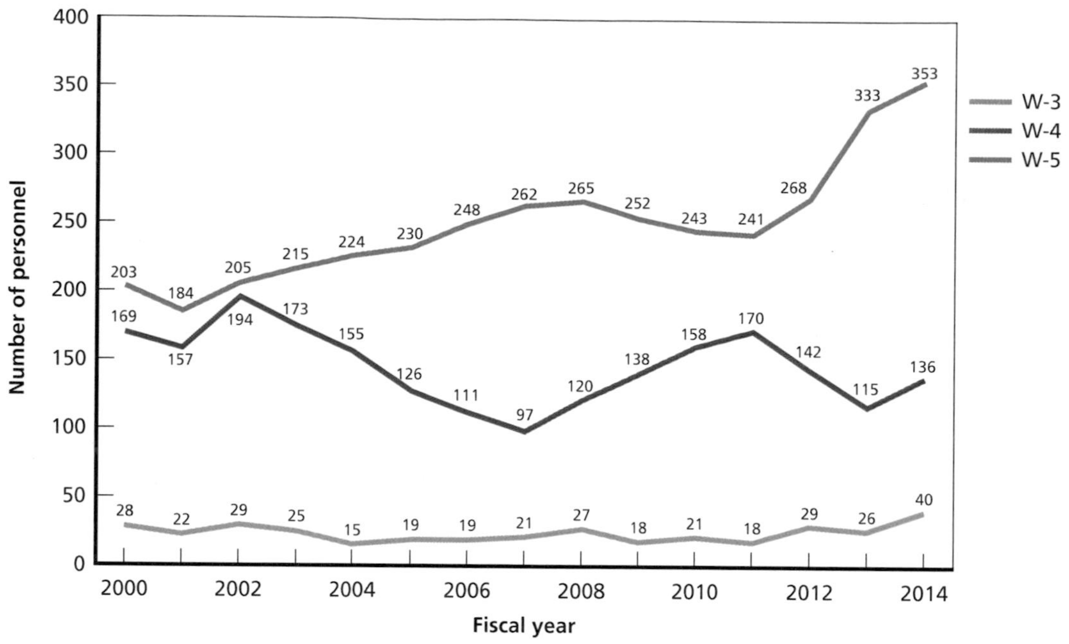

RAND *RR1209-5.8*

more than 30 years of service in FY 2005, there were 535 by FY 2014. The number of warrant officers with more than 30 years of service increased more quickly after about FY 2011. Also, there were 428 warrant officers with more than 30 years of service in 2002, but this number fell before 2005. Thus, some of the post-2007 increase merely returned the number of warrant officers with more than 30 years of service to its FY 2002 level. The overall increase may be a response to changes stemming from the recommendation of the 9th Quadrennial Review of Military Compensation.

As shown in Figure 5.8, the majority of the increase in warrant officers with more than 30 years of service occurred among W-5s and W-4s. The number of W-5s serving past YOS 30 was rising prior to 2007, from 205 in FY 2002 to 262 in FY 2007. This number then rose more sharply, to 353 by FY 2014. As noted, the increase in the number of W-5s with more than 30 years of service does not occur immediately after the legislative changes in 2007, although it could represent a delayed response as the services took advantage of the flexibility to retain more W-5s, provided by the longer pay table. That said, the number of W-4s with more than 30 years of service, in contrast, had been falling in years leading up to 2007, from 194 in FY 2002 to 111 in FY 2006. It then rose to 170 by FY 2011 before falling to 136 in FY 2014. Thus, it is possible that some of the increase in W-5s came at the expense of W-4 strength—for example, if W-5 requirements increased and were filled by faster promotion of those W-4s.

For comparison, Figure 5.9 shows the number of total warrant officers with more than 20 years of service, and Figure 5.10 shows the trend for pay grades W-3 to W-5. The number of warrant officers with more than 20 years of service increased from 5,288 in FY 2007 to 6,231 in FY 2014; however, this increase represents a continuation of the upward trend that began as early as FY 2001. Most of the increase in warrant officers with more than 20 years of service has been among W-4s and W-3s (Figure 5.10). The number of W-4s followed an

Figure 5.9
Number of Warrant Officers with More Than 20 Years of Service

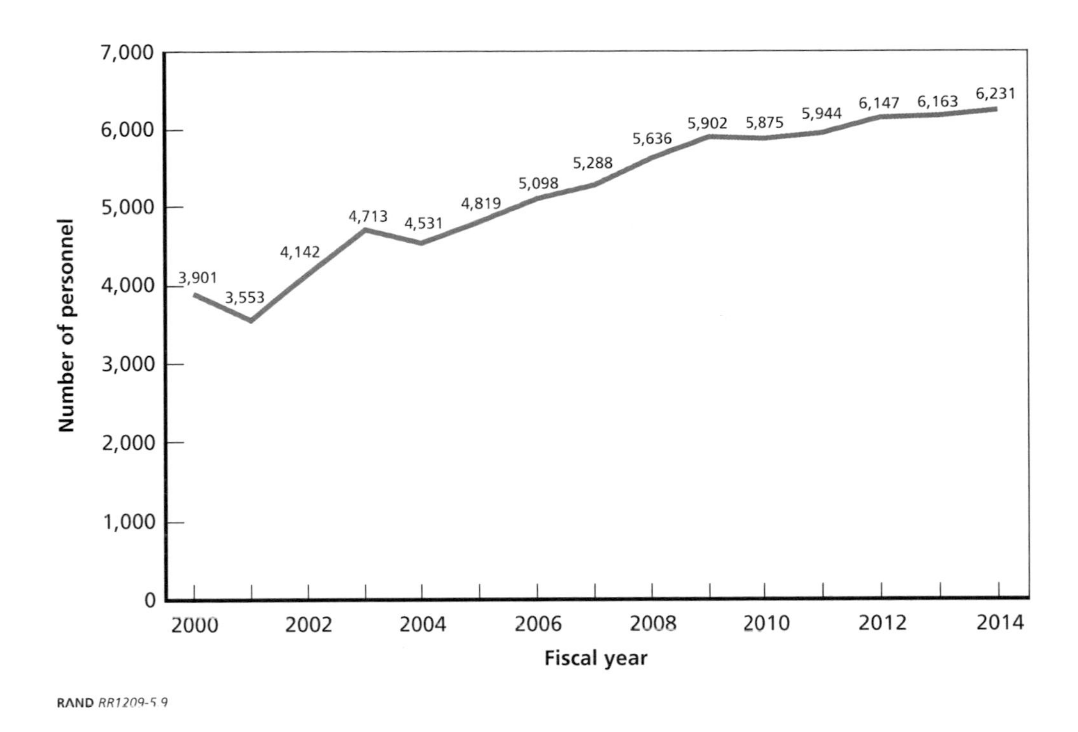

RAND *RR1209-5.9*

Figure 5.10
Number of Warrant Officers with More Than 20 Years of Service, W-3 to W-5

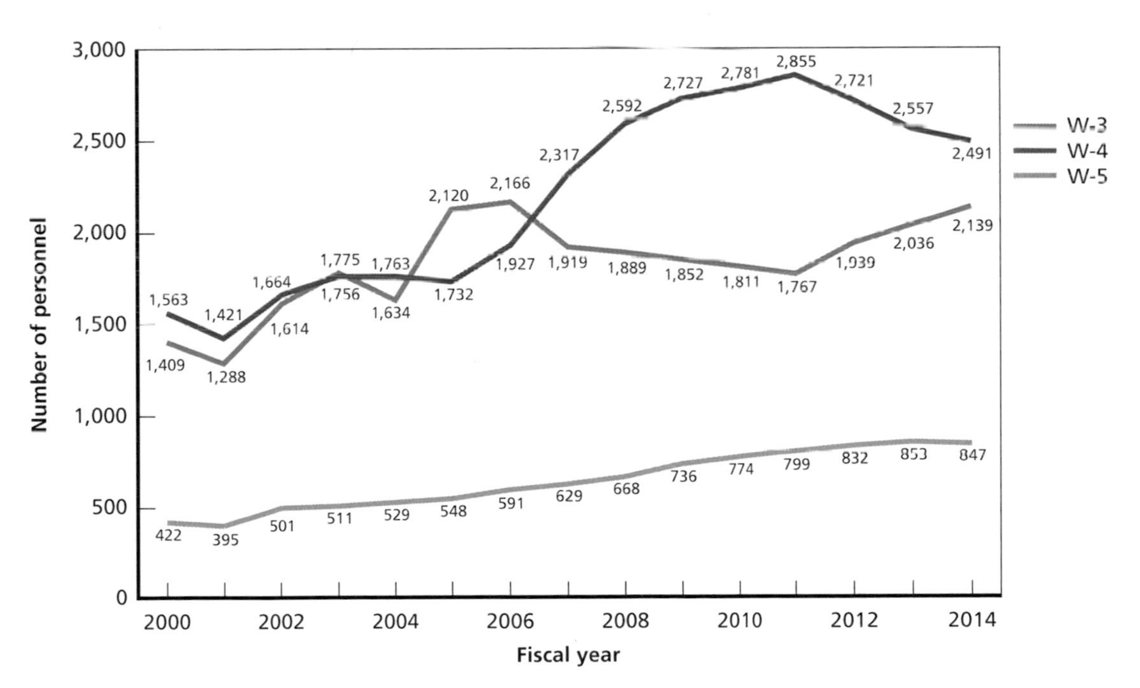

RAND *RR1209-5.10*

increasing trend from FY 2001, from 1,732 in FY 2005 to 2,855 in FY 2011, then falling to 2,491 in FY 2014. The number of W-3s with more than 20 years of service rose most rapidly prior to 2007, increasing from 1,288 in FY 2001 to 2,166 in FY 2006. This number fell to 1,767 by FY 2011 before rising to 2,139 in FY 2014. Finally, the number of W-5s with more than 20 years of service has followed a gradual upward trend since 2001, rising from 395 in FY 2001 to 847 in FY 2014.

Figure 5.11 shows the trend in number of enlisted personnel with more than 30 years of service. It also breaks out the number of E-9s with more than 30 years of service, because these personnel were a target of the pay table change. The figure shows an increase in the number of enlisted personnel with more than 30 years of service after 2005, indicating that the trend after FY 2007 is a continuation of a trend that began earlier. Focusing on the number of E-9s with more than 30 years of service, the figure shows that this number began to increase after 2004, increasing from 471 in FY 2004 to 1,386 in FY 2014.

Figure 5.12 shows a more detailed view of enlisted personnel with more than 30 years of service in pay grades E-5 through E-9. The figure shows that the number of enlisted personnel with more than 30 years of service in other grades also increased. As a point of comparison, Figure 5.13 shows the number of enlisted personnel with more than 20 years of service, which was roughly the same or even decreased since 2007 for all grades, E-5 through E-9. In fact, the peak in terms of number of personnel with more than 20 years of service appears to be between FY 2002 and FY 2004. Thus, at least for enlisted personnel, trends in the number of personnel with more than 20 and more than 30 years of service are fairly different.

Figure 5.11
Number of Enlisted Personnel with More Than 30 Years of Service

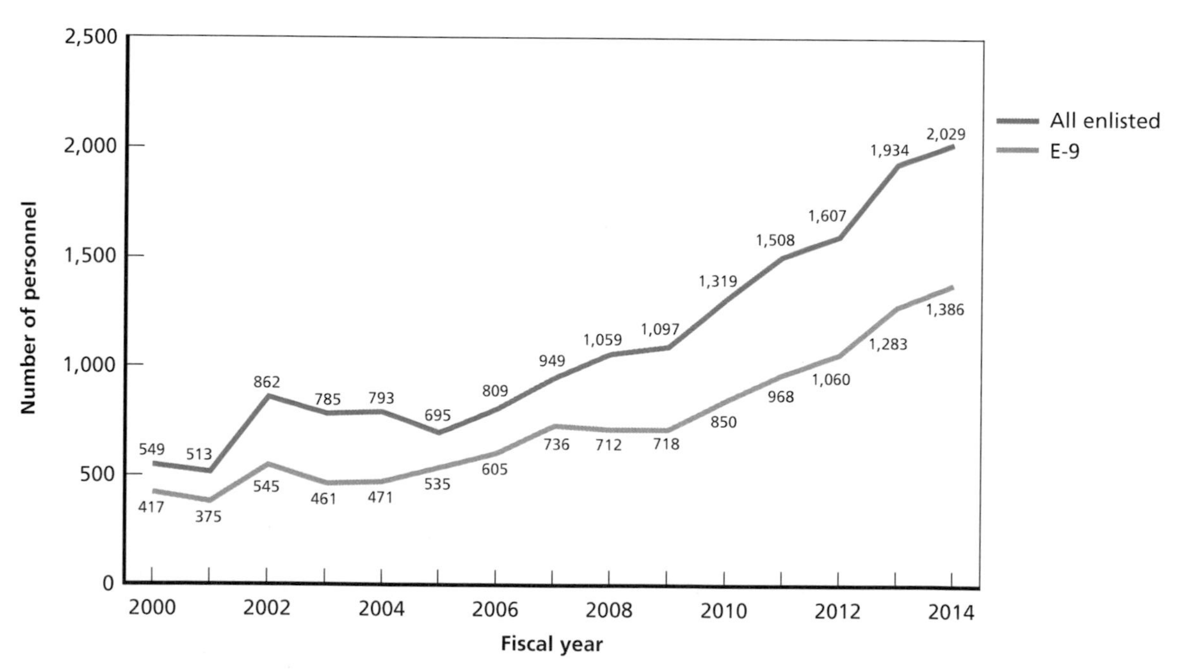

RAND *RR1209-5.11*

Figure 5.12
Number of Enlisted Personnel with More Than 30 Years of Service, by Pay Grade

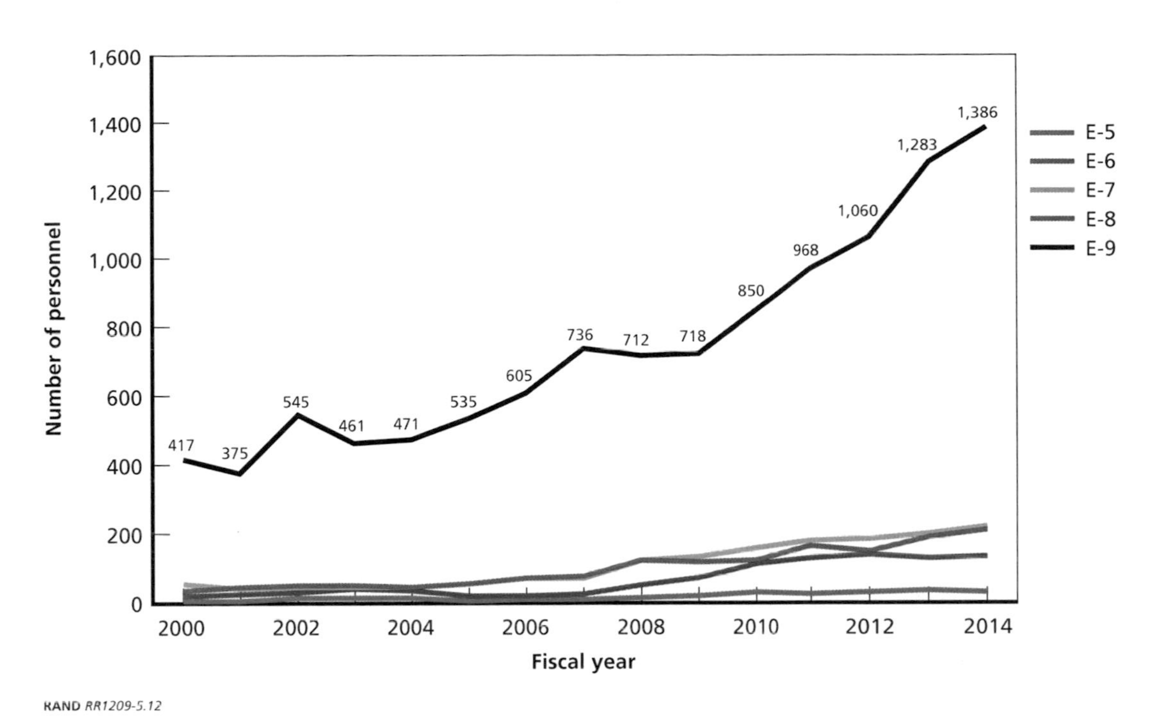

RAND *RR1209-5.12*

Figure 5.13
Number of Enlisted Personnel with More Than 20 Years of Service, by Pay Grade

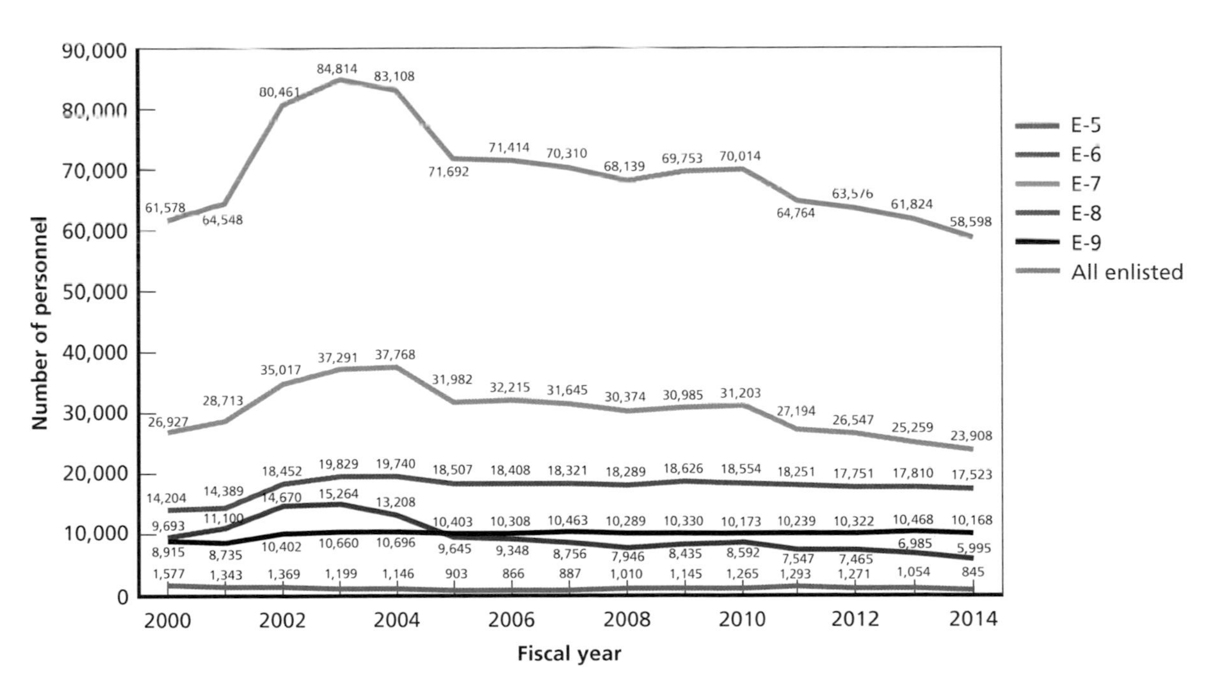

RAND *RR1209-5.13*

Tabulations by Service and Pay Grade

Though not requested by the SASC, we also produced tabulations by service and pay grade. Here, we summarize the overall trends by service, and we present the more detailed tabulations by service and pay grade in the appendix to this report. The all-service trends shown in Figures 5.1 through 5.13 hide tremendous differences across the services. We discuss each service separately.

Army

Figure 5.14 shows the number of U.S. Army officers, warrant officers, and enlisted personnel with more than 30 years of service over the period FY 2000 to FY 2014. It is clear from the figure that the number of officers and enlisted personnel with more than 30 years of service increased most drastically since 2007, with the number of warrant officers remaining more or less steady. However, it is important to note that the number of officers and enlisted personnel with more than 30 years of service was already increasing well before FY 2007, the year of the pay table change, and increased steadily starting around FY 2001. The Army faced a large demand for personnel as a result of the wars in Afghanistan and Iraq and relied heavily on emergency authorities to bring in the personnel needed. These factors no doubt explain some of the increase in the number of soldiers with more than 30 years of service.

For officers, the rate of the increase accelerated after 2007. The number of Army officers with more than 30 years of service rose from 845 in FY 2001 to 1,043 in FY 2007. It then rose to 1,747 by 2014. Similarly, the number of enlisted personnel with more than 30 years of service rose from 191 in FY 2001 to 414 by FY 2007, and to 1,313 by FY 2014. However, the number of warrant officers changed little over the FY 2000 to FY 2014 period. There were 366 warrant officers with more than 30 years of service in FY 2002 and 299 in FY 2007. This number rose to 358 by FY 2014, representing no real net change from FY 2002.

Figure 5.14
Army Personnel with More Than 30 Years of Service

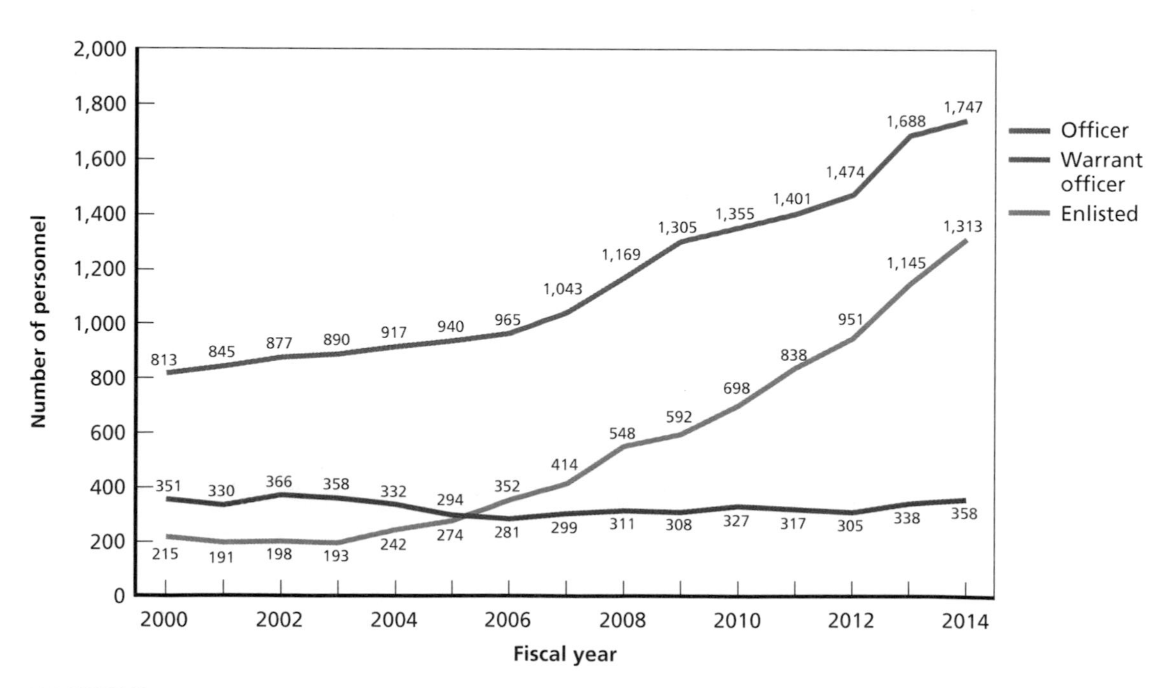

RAND *RR1209-5.14*

Air Force

Figure 5.15 shows the number of total officers and enlisted personnel with more than 30 years of service in the Air Force over the period FY 2000 to FY 2014. There are no warrant officers in the Air Force. The trends for the Air Force are different from those for the Army and for the military overall. Looking first at officers, there has been little change since 2007. The number of officers with more than 30 years of service fell in the early 2000s, from 942 in FY 2003 to 609 in FY 2007. In FY 2014, this number was 657, representing no real change over the FY 2006 level. However, looking at the number of enlisted personnel with more than 30 years of service, the figure shows an increase since FY 2007. The number was 172 in FY 2005 and rose to 310 by FY 2014.

Marine Corps

As a reminder, our trends for the Marine Corps start in FY 2002. Figure 5.16 shows the number of officers, warrant officers, and enlisted personnel with more than 30 years of service in the Marine Corps over the period FY 2002 to FY 2014. Looking first at officers, there was again little change in the number with more than 30 years of service. This number was at 243 in FY 2004 and 235 in FY 2014. The number of warrant officers with more than 30 years of service has increased since about FY 2010. This number was at 14 in FY 2006 and rose to 43 by FY 2014. This is another case where the timing of the sharpest increase does not seem to be directly related to the 2007 legislative changes. Meanwhile, the number of enlisted personnel with more than 30 years of service fell sharply, from 226 in FY 2004 to 64 in FY 2005. It then rose again to 234 in FY 2013 before falling to 160 in FY 2014. These abrupt changes could reflect data issues.

Figure 5.15
Air Force Personnel with More Than 30 Years of Service

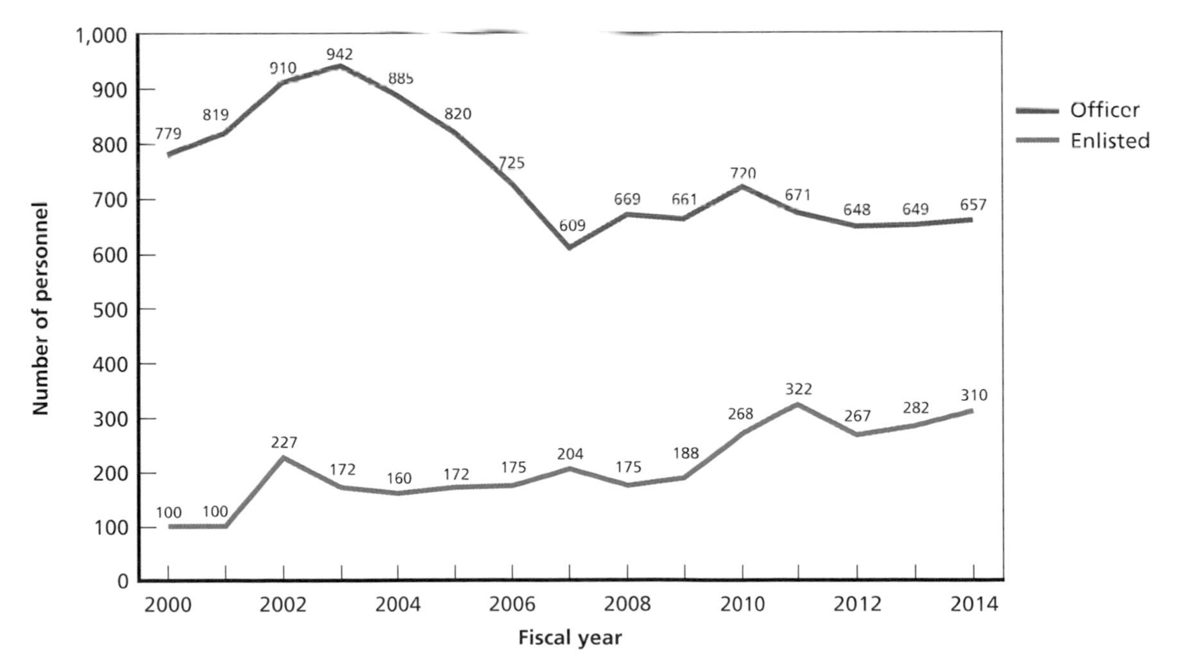

RAND *RR1209-5.15*

Figure 5.16
Marine Corps Personnel with More Than 30 Years of Service

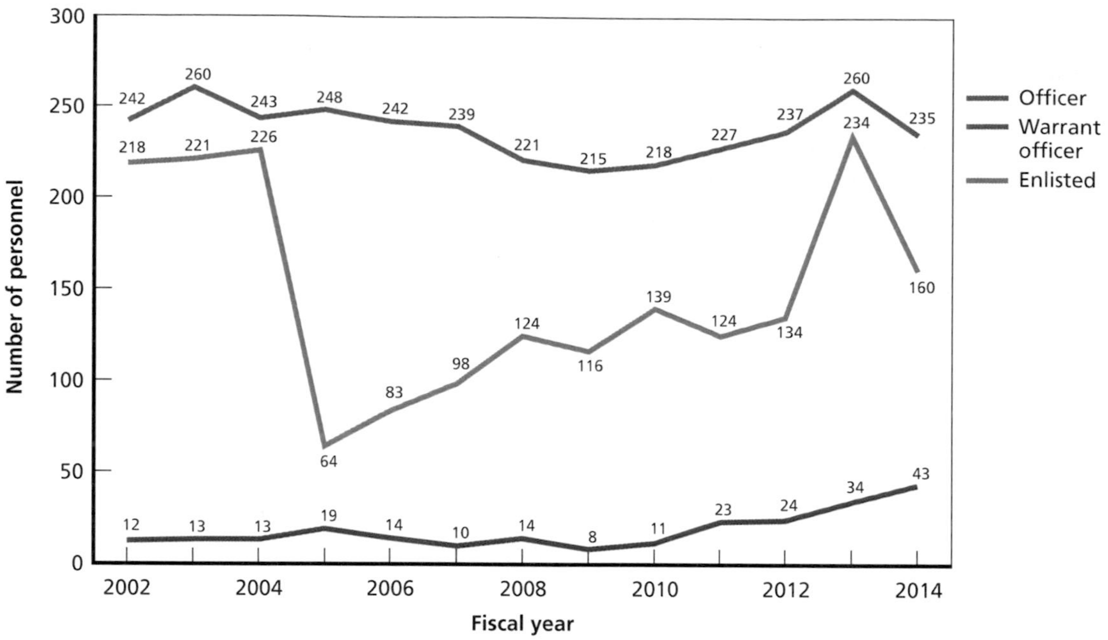

RAND *RR1209-5.16*

Navy

Figure 5.17 shows the number of officers, warrant officers, and enlisted personnel with more than 30 years of service in the Navy over the period FY 2002 to FY 2014. The increase in number of personnel with more than 30 years of service has been largest for Naval officers. This number fell between FY 2000 and FY 2004, but increased steadily afterward, from 883 in FY 2004 to 1,371 in FY 2014. As in the Army, the rate of increase was significantly faster after FY 2007 than before. The number of warrant officers has also increased since FY 2007. Again, however, this may represent the continuation of a longer trend. The number of warrant officers with more than 30 years of service rose from 49 in FY 2004 to 89 by FY 2008. This increase continued steadily to 133 in FY 2014. Finally, the number of enlisted personnel with more than 30 years of service has increased since FY 2007, but it had already been increasing starting in FY 2004. This number fell from 234 to 165 between FY 2000 and FY 2004. It then rose to 242 in FY 2014.

Analysis of Continuation Rates Across Services

In addition to considering trends in the number of senior officers and senior enlisted personnel with more than 30 years of service since 2007, we also used the Active Duty Pay file to investigate how continuation rates have changed for personnel with more than 26 years of service since 2007. Continuation rates provide information on the percentage of personnel at a given rank or year of service who choose to remain in the military in any given year. As a hypothetical example, if 40 percent of enlisted personnel with 20 years of service in 2001 remained in the service five years later, then the continuation rate to YOS 25 for enlisted personnel with 20 years of service in 2001 would be 40 percent. In our analysis, we calculated continuation rates to YOS 30 for personnel with 26 years of service from 2000 to 2011 and continuation

Figure 5.17
Navy Personnel with More Than 30 Years of Service

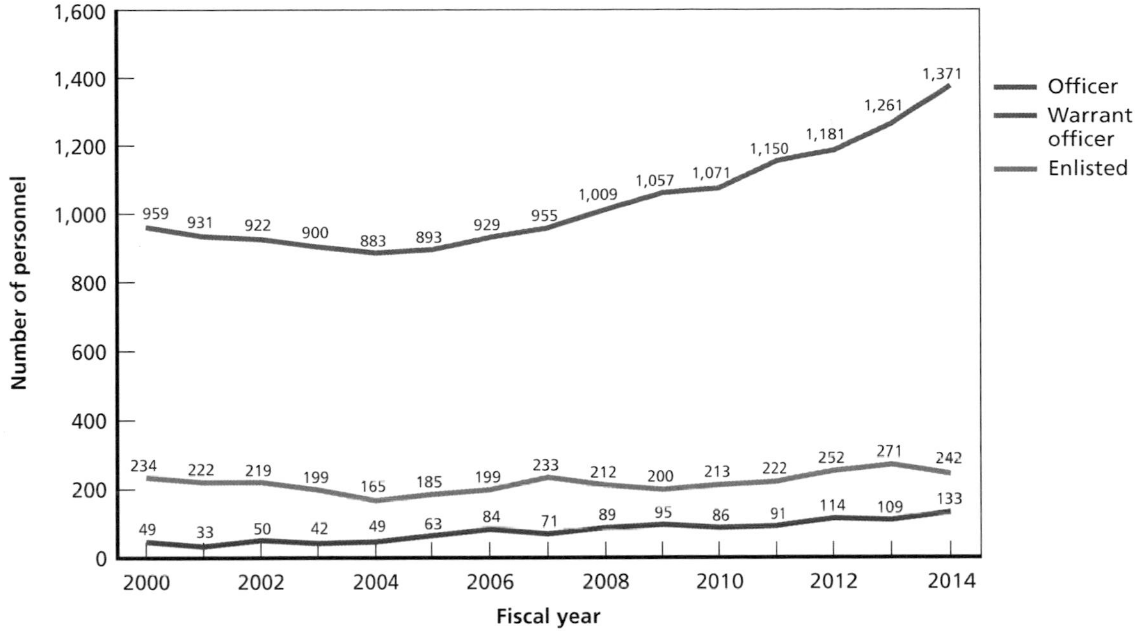

RAND RR1209-5.17

rates to YOS 32 for personnel with 30 years of service between 2000 and 2011. If the change to the 40-year pay table had the intended effect of increasing the number of senior personnel who remain in the military, then we should see an upward trend in continuation rates of personnel to YOS 30 and YOS 32.[2]

Figure 5.18 shows trends in continuation rates, considering all military services and pay grades together. The blue bars show the percentage of personnel with 26 years of service in years 2000 to 2011 who remained in the military at YOS 30. The red bars show the percentage of personnel with 30 years of service in years 2000 to 2011 who remained in the service at YOS 32. So, for example, 34 percent of personnel with 26 years of service in 2010 remained in the military at YOS 30 in 2014, while 33 percent of personnel with 30 years of service in 2010 remained in the military at YOS 32 in 2012.

The figure shows that although the number of personnel with more than 30 years of service has increased since 2007, there has been little change in the continuation rates after 2007 on net. Continuation rates rose between 2007 and 2009 among those with 26 years of service and 30 years of service, but then fell thereafter. Before 2007, continuation rates among those with 26 years of service fell, while rates among those with 30 years of service rose. On average, continuation rates before 2007 are about the same as average rates after 2007.

We can also look at continuation rates for officers, warrant officers, and enlisted personnel separately across DoD and within each service. The rest of this chapter focuses on rates across DoD for each of these subgroups. We consider continuation rates for each group by service in the appendix. As we show there, the results vary across service, but the same general pat-

[2] While the analysis of numbers of personnel with more than 20 or 30 years of service used fiscal years, our analysis of continuation rates is based on calendar years.

Figure 5.18
Continuation Rates, All Services

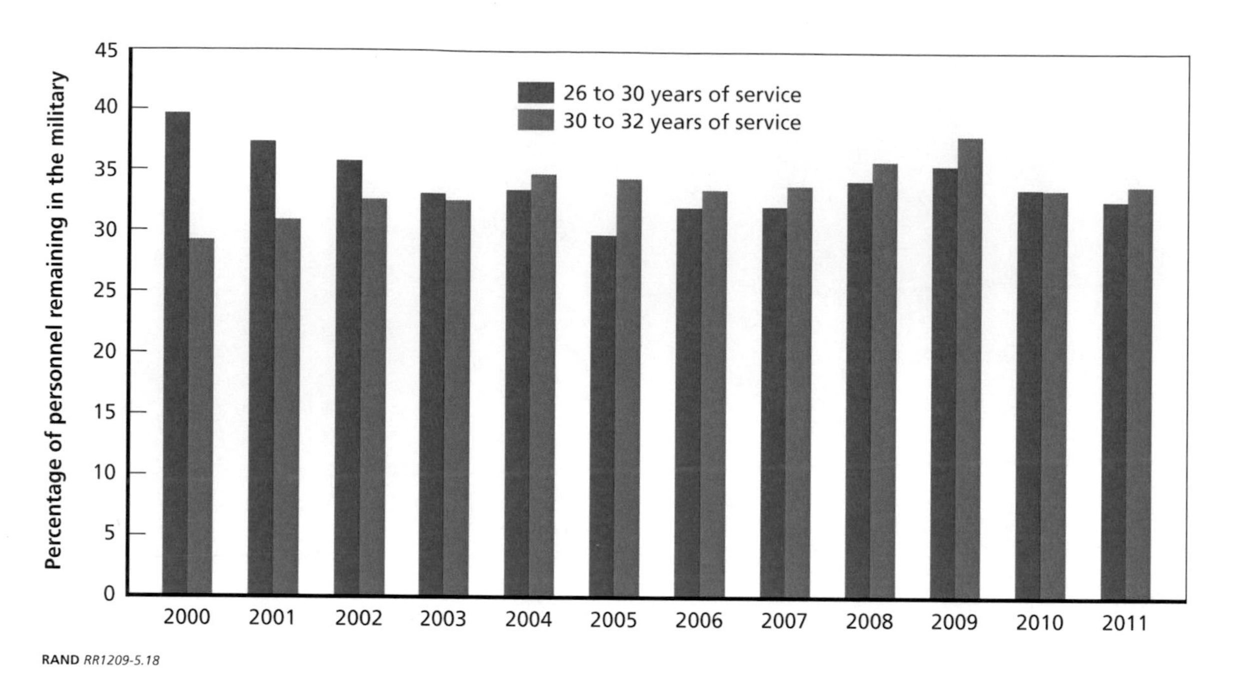

RAND *RR1209-5.18*

tern emerges, namely that continuation rates vary considerably over time but show no marked increase after 2007.

There is little change in continuation rates among officers (Figure 5.19). For personnel with 30 years of service, continuation rates to YOS 32 increased from 41 percent to 48 percent between 2000 and 2009, but fell to 45 percent in 2011, the same as the continuation rate for these individuals in 2007 and 2006. For personnel with 26 years of service, continuation rates to YOS 30 fell from 44 percent for those with 30 years of service in 2000 to 35 percent in 2005, but then rose again to 46 percent by 2009. However, this rate fell again to 40 percent in 2011. Overall, then, there is little net change and no discernible trend in officer continuation rates over the period under consideration. This is an important observation, because lengthening the careers of senior officers was a main goal of the change to a 40-year pay table.

Looking at warrant officers, there seems to be more-significant variation, but again little discernible trend in continuation rates for personnel with 26 and 30 years of service (Figure 5.20). For personnel with 26 years of service, continuation rates to YOS 30 have increased slightly on average when looking at the overall period from 2000 to 2011 and when considering the period since 2007, from 46 percent in 2000 to 48 percent in 2006 and 2007, and to 53 percent in 2011. The average continuation rate also increased slightly, from 47 percent prior to 2007 to 51 percent after 2007. For personnel with 30 years of service, continuation rates to YOS 32 have fluctuated significantly, from 27 percent for personnel with 30 years of service in 2000 to 45 percent in 2003, 29 percent in 2005, and 39 percent in 2011. In addition, the average continuation rate to YOS 32 increased slightly after 2007, from 36 percent to 43 percent. Thus, while the continuation rate for these individuals seems to have increased when comparing personnel with 30 years of service in 2000 with 2011, there does not seem to be clear evidence of a sustained upward trend since 2007 (despite the increase in the average continuation rate for this latter period) or of a direct association between changes in continuation rates for warrant officers and the change to the 40-year pay table.

Figure 5.19
Continuation Rates, Officers

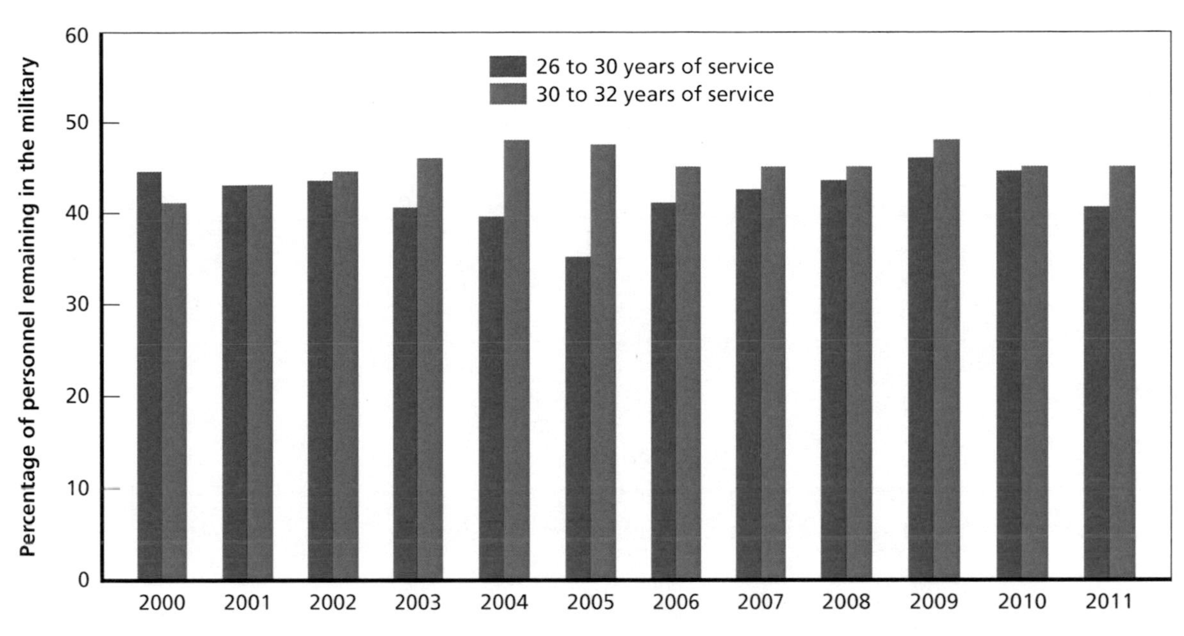

RAND *RR1209-5.19*

Figure 5.20
Continuation Rates, Warrant Officers

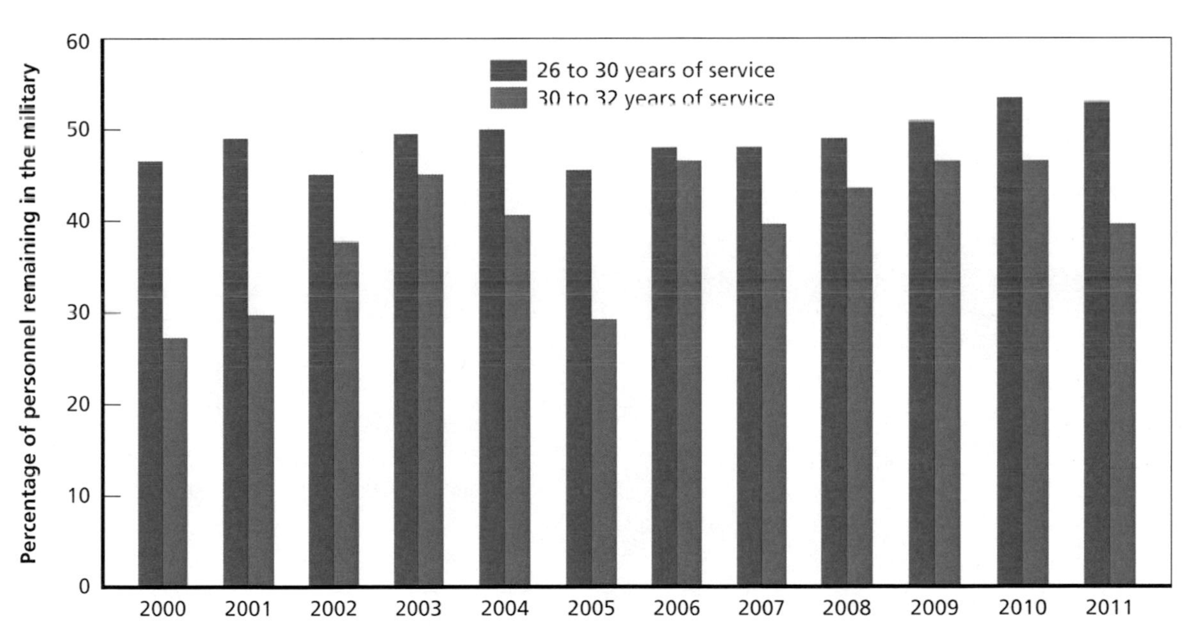

RAND *RR1209-5.20*

Figure 5.21 shows continuation rates for enlisted personnel. We see a similar pattern for enlisted personnel after 2007 as we saw in Figure 5.18. Continuation rates rose between 2007 and 2009 but fell thereafter. For those with 30 years of service, the upward trend in continuation to YOS 32 began before 2007. From 2000 to 2007, continuation rates for this group increased dramatically, nearly doubling from 13 percent to 23 percent, further increasing to 29 percent in 2009, then falling to 23 percent in 2011. The pattern differs for personnel with 26 years of service. Continuation rates to YOS 30 for personnel with 26 years of service decreased between 2000 and 2005 before increasing to 30 percent in 2009, and then falling again to 27 percent in 2011, for an overall decrease between 2000 and 2011 of around 7 percent.

Summary

Taken together, our tabulations offer some valuable insights into the potential effects of the change to the 40-year pay table in 2007. First, the overall number of personnel with more than 30 years of service has increased since 2007, especially for the Army (but also in the other services). Although there were differences across services, these increases were greatest among senior enlisted personnel (specifically E-9s) and warrant officers, who were a secondary target of the change to the 40-year pay table. For officers, the greatest increase in personnel with more than 30 years of service appears to be among O-5s and O-6s, possibly individuals with prior enlisted service or recalled retirees, as opposed to flag grade officers (O-7 to O-10), who were the primary target of the change to the 40-year pay table. In addition, the upward trends that we observed in the number of personnel with more than 30 years of service often begin well before

Figure 5.21
Continuation Rates, Enlisted Personnel

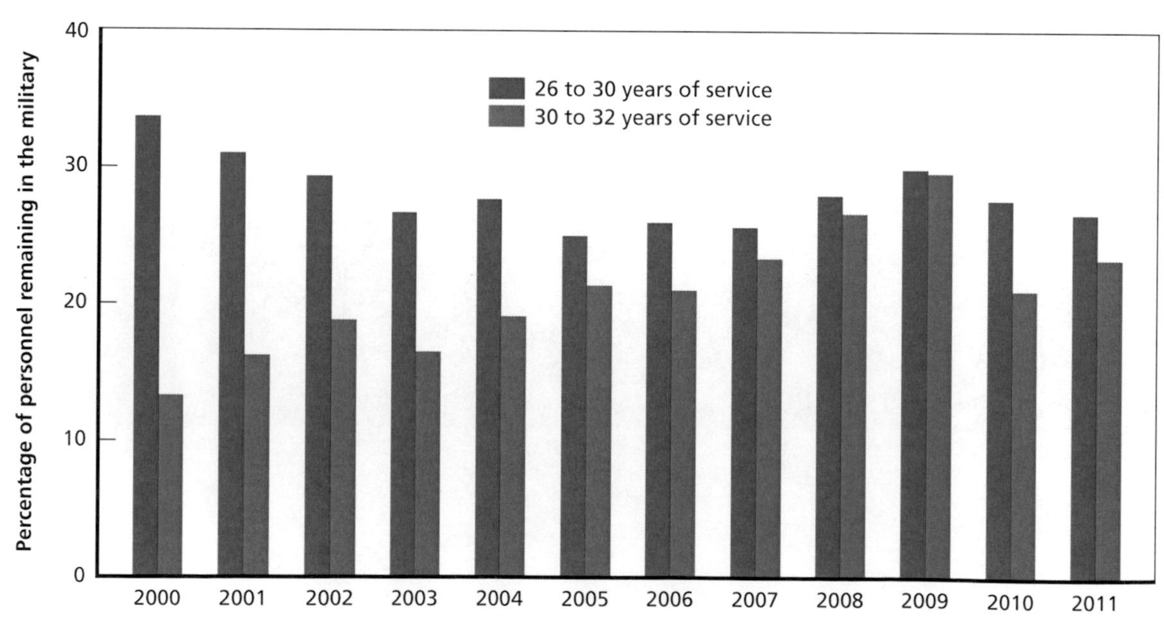

RAND *RR1209-5.21*

(and in some cases well after) 2007 and may be tied to the increased demands placed on the military during the Afghanistan and Iraq wars, the global recession, and other contextual factors.

However, despite the apparent increase in the number of personnel with more than 30 years of service, our analysis of continuation rates did not reveal a significant increase in retention rates for senior personnel after 2007. This was true across services and pay grades, with a few exceptions. We did observe some evidence of an increase in continuation rates for warrant officers, Army enlisted personnel with 30 years of service (to YOS 32), Marine Corps enlisted personnel with 26 years of service (to YOS 30), and Marine Corps warrant officers at 26 and 30 years of service (when focusing on average rates pre- and post-2007). However, offsetting these small upward trends in the continuation rates are several instances of downward trends, specifically in the Air Force. Furthermore, in all cases, upward trends are modest in size and very often do not appear directly linked to 2007 legislative changes, beginning before 2007 or occurring after 2007. Put differently, we observe considerable variation in continuation rates but little evidence of a marked increase in rates after 2007.

Considering the lack of change in continuation rates alongside the observed increase in the number of personnel serving past 30 years suggests that those increases in personnel with more than 30 years of service may have been concentrated among very specific groups of people—for example, officers with prior enlisted experience who stayed for an extra assignment, senior enlisted and warrant officers who similarly were retained to fill specific jobs, and recalled retirees who returned to support the increased pace of deployment. Furthermore, the lack of a marked change in rates of continuation after 2007, despite the increase in the number of personnel with more than 30 years of service, suggests that the increase in the number of senior personnel may be at least in part a function of an increase in the overall size of the military in the period since 2001, driven by increased demands and pace of deployment in this period, rather than a true increase in the percentage of personnel choosing to stay in the military past 30 years. If true, this would be consistent with the perception of our interviewees that the legislative changes in 2007 had little effect on the retention decisions of the majority of service members.

CHAPTER SIX

Simulated Retention Effects from the Dynamic Retention Model

The DRM is well suited to analyzing the retention effects of the 2007 pay table changes and of stopping longevity increases beyond YOS 26, thereby reverting to a 30-year table for senior personnel. The model is designed to address questions related to how changing the level and structure of compensation affects retention over a military career, as well as cost in the steady state and in the transition to the steady state. As discussed in Chapter Five, trends in retention and force size before and after 2007 are difficult to interpret as being causal evidence because of other factors affecting observed trends. These factors include the effects of the recession that began in late 2008, changes in wartime requirements, and changes in service personnel management policies. The DRM provides a causal assessment of the retention and cost effects of the pay table changes, abstracting from the effects of these other factors. This chapter summarizes the results of our DRM analysis.

We use the DRM to simulate the retention effects of the following policy changes:

1. the four changes that went into effect in 2007, summarized in Chapter One:[1]
   a. moving from the 30-year to the 40-year pay table
   b. increasing the cap on basic pay from Executive Level III to Level II
   c. removing the cap on basic pay for computing retired pay
   d. removing the cap on YOS for computing the retirement benefit multiplier[2]
2. the reinstatement in 2014 of the cap on basic pay for computing retired pay
3. the reinstatement of a 30-year pay table, as discussed by the SASC report
4. the reinstatement of a 30-year pay table together with a special and incentive pay for senior personnel to sustain retention relative to the 40-year table.

The first two simulations show the effects of actual policy changes, the first occurring in 2007 and the second in 2014 as part of the 2015 NDAA. The second two simulations show the effects of proposed policies, one involving a change back to the 30-year pay table, but retaining existing policy with respect to caps on basic pay and the retired pay multiplier, and one that involves reinstating the 30-year table but also using special and incentive pays to sustain retention of experienced personnel. These last two simulations demonstrate whether the retention of experienced personnel could be equally achieved with a 30-year table, the extent to which special and incentive pays would be needed to achieve that retention, and at what cost.

[1] As mentioned in Chapter One, the change to the 40-year table occurred in April 2007, not January 2007. The increase of the basic pay cap from Executive Level III to Level II occurred in January 2007. Public Law 109-364, enacted in October 2006, eliminated the 75-percent multiplier cap for retired pay for those retiring beginning in January 2007.

[2] Our simulations model retention over a 40-year career, so the maximum multiplier in our simulations is 100 percent.

Before presenting the simulation results, we provide a brief overview of the model. The DRM has been documented in detail in a number of previous publications,[3] and interested readers are directed to those reports for further information.

DRM Overview

As described in Chapter Three, the foundation of the DRM is a theory of retention decision-making over a service member's career. The theory is a mathematical model of individual decisionmaking in a world with uncertainty and in which individuals are heterogeneous in terms of their tastes for military service. The model begins with service in the active component, and individuals make a decision to stay or leave each year. Those who leave the active component take a civilian job and, at the same time, choose whether to participate in the reserve component. That decision is made each year, and the individual can move into or out of the reserve component from year to year. More specifically, a reservist can choose to remain in the reserves or to leave it to be a "civilian," and a civilian can choose to enter the reserve component or remain a civilian.

The parameters of this model are empirically estimated, with data on military careers drawn from administrative data files, specifically the Defense Manpower Data Center Work Experience file. The file contains person-specific longitudinal records of active and reserve service. We used the data for service members who began their military service in 1990 and 1991 and tracked their individual careers in the active component, and, if they joined, the reserve component, through 2010, providing 21 years of data on 1990 entrants and 20 years on 1991 entrants. For each active component, we drew samples of 25,000 individuals who entered the component in FY 1990–1991, constructed each service member's history of active and reserve component participation, and used these records in estimating the model. We supplemented these data with information on active, reserve, and civilian pay. Active-component pay, reserve-component pay, and civilian pay are averages based on the individual's years of active-component experience, active- plus reserve-component experience, and total experience, respectively. We used 2007 military pay tables, but because military pay tables have been fairly stable over time, with few changes to their structure,[4] we did not expect our results to be sensitive to the choice of year. For civilian pay opportunities for enlisted personnel, we used the 2007 median wage for full-time male workers with an associate's degree. For officers, we used the 2007 80th-percentile wage for full-time male master's degree holders in management occupations. The data were from the U.S. Census Bureau.

Because observed data from the Work Experience file do not extend to longer careers and, in particular, do not include 30 or more years of service, we took additional measures to ensure that the percentage of personnel with more than 30 years of service closely matched the actual

[3] Mattock, Hosek, and Asch, 2012; Asch, Hosek, and Mattock, 2012, 2014; and Asch, Mattock, and Hosek, 2013.

[4] An exception was the structural adjustment to the basic pay table in FY 2000 that gave larger increases to mid-career personnel who had reached their pay grades relatively quickly (after fewer years of service). A second exception was the expansion of the basic allowance for housing, which increased in real value between FY 2000 and FY 2005.

percentage. This was important because our DRM analysis focused on the retention behavior of senior personnel. We approximated the actual percentage from counts of personnel entering the active component before 1990–1991; such counts were used for the tabulations reported in Chapter Five. Given the actual percentage, we then coded the simulation of the baseline (current) system to allow members to look forward and anticipate that only some percentage of those who were willing to continue past 30 years of service would be permitted to do so, and further, when simulated personnel reached 30 years of service, they were subject to a selection process that kept only some percentage of them. This approach combined two aspects of the intuition in the dynamic retention approach—namely, forward-looking behavior and the realization of shocks in the future period. The anticipated rate of being allowed to continue past 30 years factored into forward-looking behavior, while the actual selection at 30 years was a realized selection. This approach allowed us to ensure that, at baseline, the percentage serving beyond 30 years was close to the actual value. Then, when we simulated an alternative policy, such as reverting from the 40-year to the 30-year pay table, we assumed that the anticipated and realized selection rates were the same.

A limitation of this model is that the Army National Guard and Army Reserve were not treated separately but were combined into a single group, the Army Reserve component. The Air Force Reserve and Air Guard were also combined into a single group. The model assumed that military pay, promotion policy, and civilian pay were time stationary, and we excluded demographic variables, such as gender, marriage, and spousal employment. We also excluded health status and health care benefits, and we did not explicitly model deployment or deployment-related pays. That said, the estimated models fit the observed data extremely well for the both the active and reserve components.

Thus, in short, the DRM is firmly grounded in the theory of retention decisionmaking and empirically grounded in data on the actual retention behavior of thousands of service members over a 20-year period. Further, the DRM includes a simulation capability that allows assessment of major compensation reforms without relying on the existence of prior variations in such reforms. Because the model is formulated in terms of the parameters that underlie the retention decision process rather than on members' average response to a particular compensation policy, the capability is structured to enable assessments of alternative compensation systems that have yet to be tried or of policy changes for which it would be difficult to get a counterfactual of what would have happened in the absence of the policy change. That is, it permits "what if" analyses of changes in compensation policy, even for changes that may be outside of historical experience, such as a change in retirement compensation.

We have model estimates and simulation capability for each service and within each service for officers and for enlisted personnel, and for the active component and for the reserve component, conditional on prior active-component service. In our past work (Mattock, Hosek, and Asch, 2012; Asch, Hosek, and Mattock, 2013, 2014), we have found that the retention results and the percentage changes in costs for the Army are representative of those for the other services, in terms of magnitude of effects. For the purpose of costing, we conduct the analyses for all of the services, but to illustrate results, we show simulation outcomes for the Army only. We find that the simulated retention results for the other services are quite similar to those found for the Army.

Effects of the 2007 Policy Changes

Figure 6.1 shows the retention effects of the 2007 policy changes for Army officers, and Figure 6.2 shows the effects for Army enlisted personnel. The left panel in each figure shows the cumulative probability of an entrant being retained at each YOS, from entry to YOS 40. For example, at entry (YOS 0), the cumulative probability is 1. Because of the relatively small number of personnel who serve beyond YOS 30, the right panel in each figure zooms in on YOS 30–40. That is, it shows the probability that an entrant reaches YOS 30 to YOS 40, shown on a different scale to highlight differences.

The black line in each graph shows the baseline retention prior to the 2007 changes. That is, it shows the retention profiles of the Army officer and enlisted forces under the pre-2007 environment—namely, the 30-year pay table, the Executive Level III cap for basic pay (as well as for basic pay for the purpose of computing retired pay), and the 75-percent retirement multiplier cap. The note at the bottom of each figure shows the number of personnel with more than 30 years of service as a percentage of those who have served more than 20 years. In the baseline, 7.65 percent of officers with more than 20 years of service have served more than 30 years, while the figure is 1.38 percent for Army enlisted personnel.

The red line in each graph shows the simulated effect of the 2007 policy changes. It shows the effect of extending the pay table to 40 years, raising the basic pay cap to Executive Level II pay, removing the cap on basic pay for the purpose of computing retired pay, and lifting the cap on the retired pay multiplier so that it is 100 percent at YOS 40. The simulated retention profiles assume that the Army made no change to personnel policy with respect to tightening or loosening its rules on who is permitted to stay; the simulations that produce the profiles also abstract from other changes that occurred, such as changes in the civilian economy or wartime requirements.

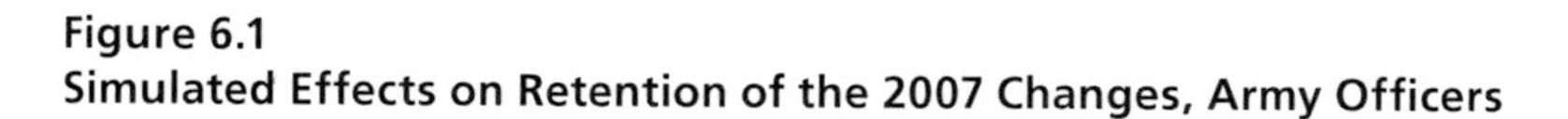

Figure 6.1
Simulated Effects on Retention of the 2007 Changes, Army Officers

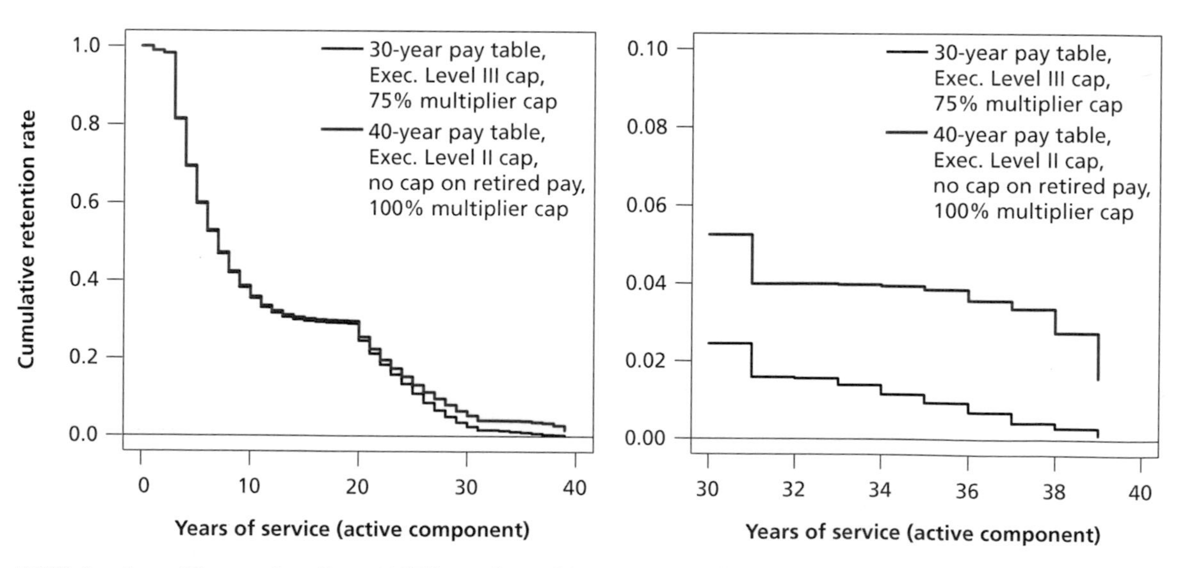

NOTE: Serving > 30 years, baseline = 7.65%; serving > 30 years, new policy = 19.59%.

RAND *RR1209-6.1*

Figure 6.2
Simulated Effects on Retention of the 2007 Changes, Army Enlisted Personnel

NOTE: Serving > 30 years, baseline = 1.38%; serving > 30 years, new policy = 2.66%.
RAND *RR1209-6.2*

It is clear from the graphs that the changes that occurred in 2007 increased steady-state retention for officers, and, to a smaller extent, for enlisted personnel. Importantly, because members are forward-looking, retention increased not just among personnel with more than 30 years of service but also among those with more than 20 years. That is, while the policy changes directly affect the compensation of those with more than 26 years of service, more-junior members anticipate those changes, affecting their current retention decisions. The retention effect grows with years of service. We find that the number of personnel with more than 30 years of service as a percentage of those with more than 20 years of service more than doubles, from 7.65 percent to 19.59 percent, for officers, and it increases from 1.38 percent to 2.66 percent for enlisted personnel. As seen in the right panel of Figure 6.1, the retention effect continues to grow even after YOS 30 for officers; retention is relatively flat for officers between YOS 31 and YOS 35 under the 2007 changes but declines in the pre-2007 regime. That is, those who would have left in the pre-2007 pay table regime continue to stay. For enlisted personnel, the largest increase in retention after YOS 30 is in the earlier years, between YOS 30 and 34.

For enlisted members, extending the pay table to 40 years adds longevity increases for senior members who served beyond YOS 26, thereby increasing retention incentives. Furthermore, lifting the cap on years of service for computing the retirement multiplier means that members serving beyond YOS 30 accrue additional retirement earnings, equal to 2.5 percent of basic pay for each additional year. This also increases retention incentives. Thus, the increase in enlisted retention after YOS 30 is consistent with what one would expect.

Officer retention incentives also increase because of the lifting of the cap on the retired pay multiplier. In addition, the increase in the cap from Executive Level III to Level II raises the basic pay allowable to the most-senior officers, further increasing the incentive to stay in service. Also, removing the Executive Level cap for computing retired pay means that each additional year of service will result in additional retirement earnings for officers, not just through the higher multiplier. Because the Executive Level cap is relevant only to officers, we expect the retention effects of the 2007 changes to be larger than they are for enlisted personnel, and that is indeed the case.

These results imply that the 2007 changes increase retention for those with more than 20 years of service, inducing longer careers among these personnel, all else equal. Interestingly, among those with fewer than 20 years of service, we see no change in retention for Army officers and a slight increase for Army enlisted personnel. While the military retirement system continues to be a key driver of retention among mid-career personnel with fewer than 20 years of service for both officers and enlisted personnel, the changes in 2007 affecting current compensation have a larger effect for mid-career enlisted personnel than for mid-career officers.

Effects of the 2014 Policy Change

In the 2015 NDAA, Congress restored the Executive Level II cap on basic pay for the purpose of computing retired pay for years served after December 2014. Figure 6.3 shows the retention effect for Army officers of this change. We do not show results for enlisted personnel, because the cap—and therefore the 2014 change—has no effect on basic pay for these members, as seen in Chapter Two.

As for officers, restoring the cap has virtually no effect on retaining officers with fewer than 34 years of service. However, Army officer retention falls among those with 34 or more years of service. With the policy change, additional service increases retired pay only by increasing years of service and not by increasing basic pay, as was the case before the 2014 change. Thus, each year served beyond 30 represents a year of forgone retired benefits. By restoring the cap for computing retired pay, only the increase in the retired pay multiplier is available to offset this loss. We find that the number of officers with more than 30 years of service as a percentage of those with more than 20 years falls from 19.59 percent to 17.17 percent for Army officers.

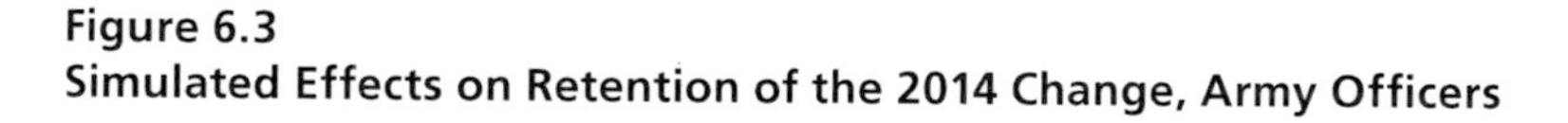

Figure 6.3
Simulated Effects on Retention of the 2014 Change, Army Officers

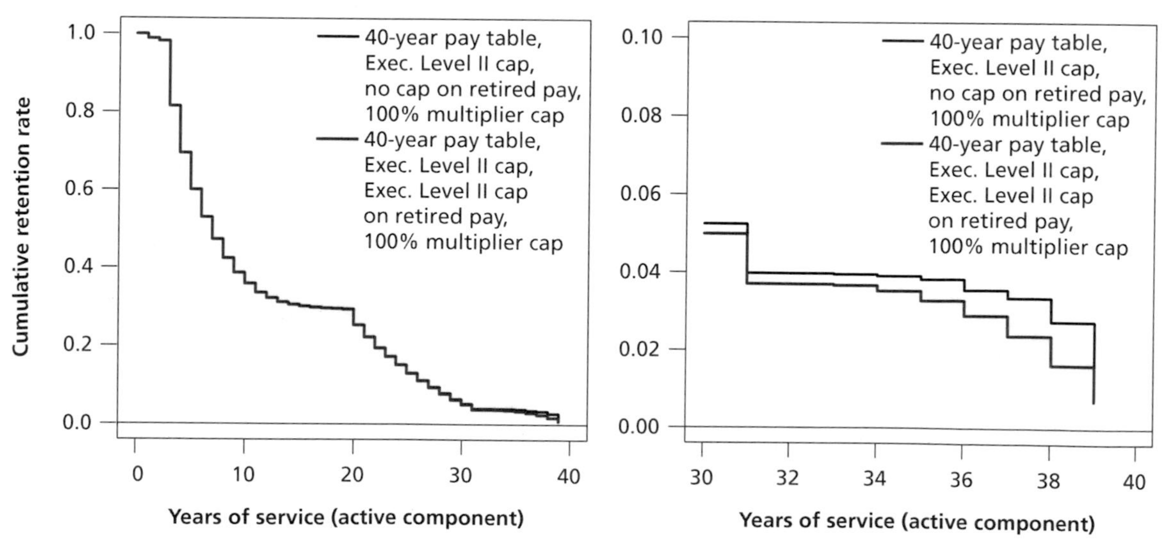

NOTE: Serving > 30 years, baseline = 19.59%; serving > 30 years, new policy = 17.17%.

RAND *RR1209-6.3*

Effects of Reverting to the 30-Year Table

The SASC report asked whether the retention of experienced personnel who would otherwise be difficult to retain would be achievable under the 30-year pay table. We address that question in this subsection. We first show the retention effects for Army officers and enlisted personnel of changing the pay table to one where no longevity increases occur for service beyond YOS 26 relative to the 40-year table, but no other change occurs. That is, the simulations continue to hold the cap on basic pay at Executive Level II (for both basic pay and computing retired pay) and do not impose a retirement multiplier cap at 75 percent. We then show whether the use of a special and incentive pay, targeted at YOS 30, can sustain retention in the context of a 30-year table. That is, we consider whether and how much continuation pay would be needed to achieve the same retention profile under the 30-year table as under the 40-year table.

Figure 6.4 shows the simulated results of reverting to the 30-year table, with no other change, for Army officers. Figure 6.5 shows the simulated results for Army enlisted personnel. Retention falls among personnel not only with more than 26 years of service but also with 20–26 years of service, though the effect is less easily seen for enlisted personnel, given the scale of the graphic. Members with less than 26 years of service anticipate the lack of longevity increases beyond YOS 26 and are consequently less likely to stay. The drop in retention among those with more than 30 years of service is seen in the right panels of Figures 6.4 and 6.5. For Army officers, the number of personnel with more than 30 years of service as a percentage of those with more than 20 years of service falls from 17.17 percent to 13.53 percent; it falls from 2.66 percent to 1.52 percent for Army enlisted personnel.

Whether the drop in retention after YOS 30 poses a problem depends on the services' requirements. If the requirements for the most-experienced personnel are lower, say as a result of a military drawdown, then the drop in retention could help achieve a lower requirement. Alternatively, even if requirements are not lower, it is possible that the services could loosen

Figure 6.4
Simulated Effects on Retention of Reverting to the 30-Year Table, Army Officers

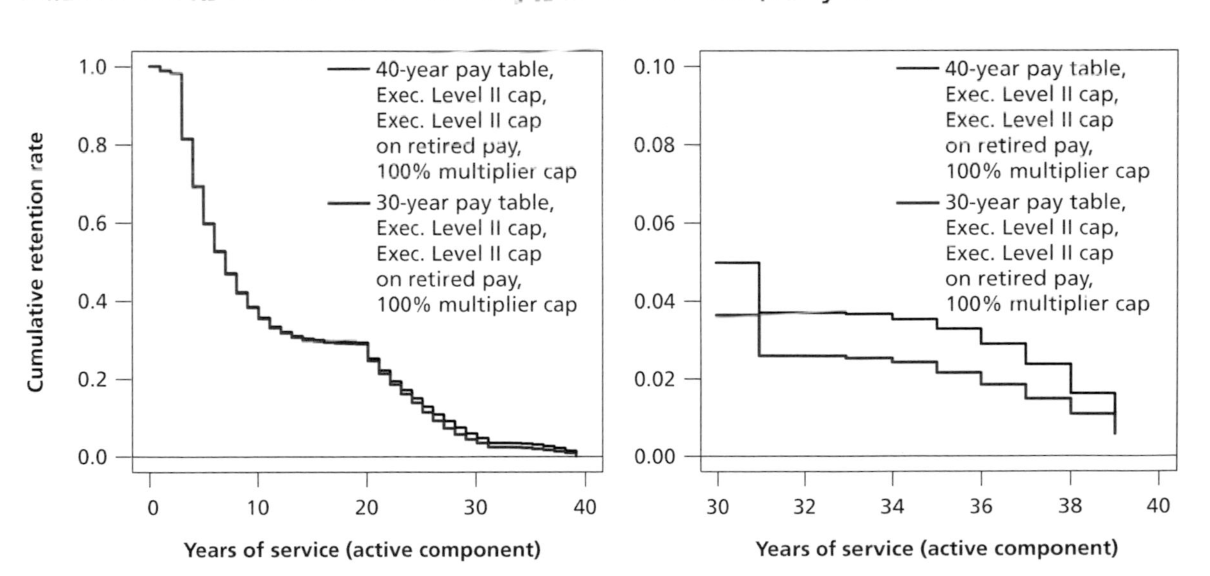

NOTE: Serving > 30 years, baseline = 17.17%; serving > 30 years, new policy = 13.53%.
RAND *RR1209-6.4*

Figure 6.5
Simulated Effects on Retention of Reverting to the 30-Year Table, Army Enlisted Personnel

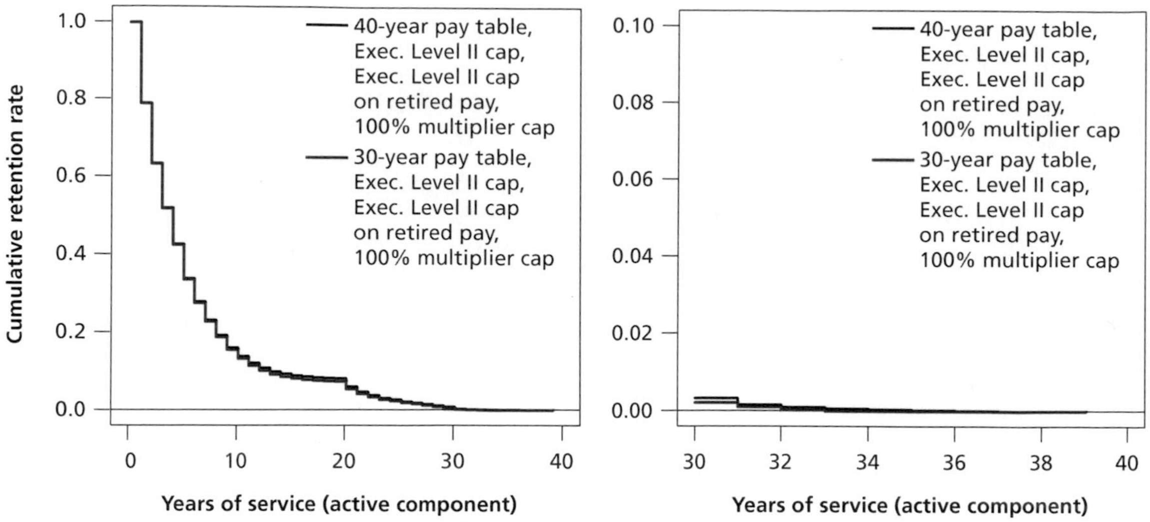

NOTE: Serving > 30 years, baseline = 2.66%; serving > 30 years, new policy = 1.52%.
RAND *RR1209-6.5*

personnel policies that induce mandatory separation, thereby allowing more personnel to stay. That is, the drop in retention might be manageable with other policy tools.

Still, the reduction in retention shown in Figures 6.4 and 6.5 might be sufficiently large that the use of additional compensation is required. Therefore, we considered how a special and incentive pay might be used to sustain retention under a 30-year table relative to a 40-year table. Specifically, we considered a lump sum cash payment paid to members at YOS 30. That is, the payment would be made for completion of YOS 29 and paid at the beginning of YOS 30,[5] regardless of whether a member subsequently stayed in service beyond that point. Thus, the special and incentive pay is like a gate pay; members receive it for completing a particular milestone—namely, reaching YOS 30. In our analysis, all members reaching this milestone are paid the special and incentive pay, though in reality, the services might consider targeting the pay to specific members and changing the milestone to a different year, such as YOS 32.

The question of interest is how large a special and incentive pay needs to be, and how much would need to be budgeted, to restore retention under a 30-year table to that under a 40-year table. That is, what amount of pay is needed to minimize the gap between the black and red lines in Figures 6.4 and 6.5? We developed an optimization routine in the DRM coding that finds the optimal amount of the special and incentive pay at YOS 30 to minimize the distance between the red and black lines. We find that for Army officers, the amount is $99,600, in 2015 dollars. For the other services, we find that the amount for officers is a bit lower: $87,900 for Marine Corps officers, $90,900 for Air Force officers, and $92,000 for Navy officers. We find for Army enlisted personnel, the amount is $37,400 in 2015 dollars. This is lower than what we find for enlisted personnel in the other services: $41,700 for the Marine Corps, $51,200 for the Air Force, and $58,200 for the Navy.

[5] Our YOS numbering begins with year 0, so in YOS 29, the member is serving in his or her 30th year.

The retention profile under the 30-year table and the special and incentive pay compared with the profile under the 40-year table is shown in Figure 6.6 for Army officers and in Figure 6.7 for Army enlisted personnel. The figures show that the special pay induces members with fewer than 30 years of service who would have left in the absence of the special pay to remain in service to receive the incentive pay. For officers, given that individuals stay longer to YOS 30 when offered special pay, we find that there remains little difference in retention after YOS 30 under the 30-year versus 40-year pay table, with a slight increase just at YOS 30. Thus, under the 40-year table, 17.17 percent of Army officers with more than 20 years of service have served more than 30 years (given the Executive Level II caps), while under the 30-year table with the optimized payment of $99,600, the percentage is nearly the same, at 17.33 percent. Furthermore, retention is nearly the same not only after YOS 30 but also before YOS 30. Thus, the special pay is sufficient inducement to buy back officer retention.

The results for enlisted personnel are generally similar. The special pay improves retention after YOS 30. In fact, if anything, retention after YOS 30 is actually a bit higher under the 30-year table with special pay than under the 40-year table; the number of enlisted personnel with more than 30 years of service as a percentage of those with more than 20 years increases from 2.66 percent to 2.89 percent. On the other hand, retention among those with less than 20 years of service continues to be a little bit lower than retention under the 40-year table, even with the addition of the special pay.

While not shown, the results for the other services are quite similar qualitatively. For officers, retention is nearly identical before and after YOS 30 under the 30-year versus 40-year table, given the addition of the special pay under the 30-year table. For enlisted personnel, spe-

Figure 6.6
Simulated Effects on Retention of Reverting to the 30-Year Table and Adding Special Pay at YOS 30, Army Officers

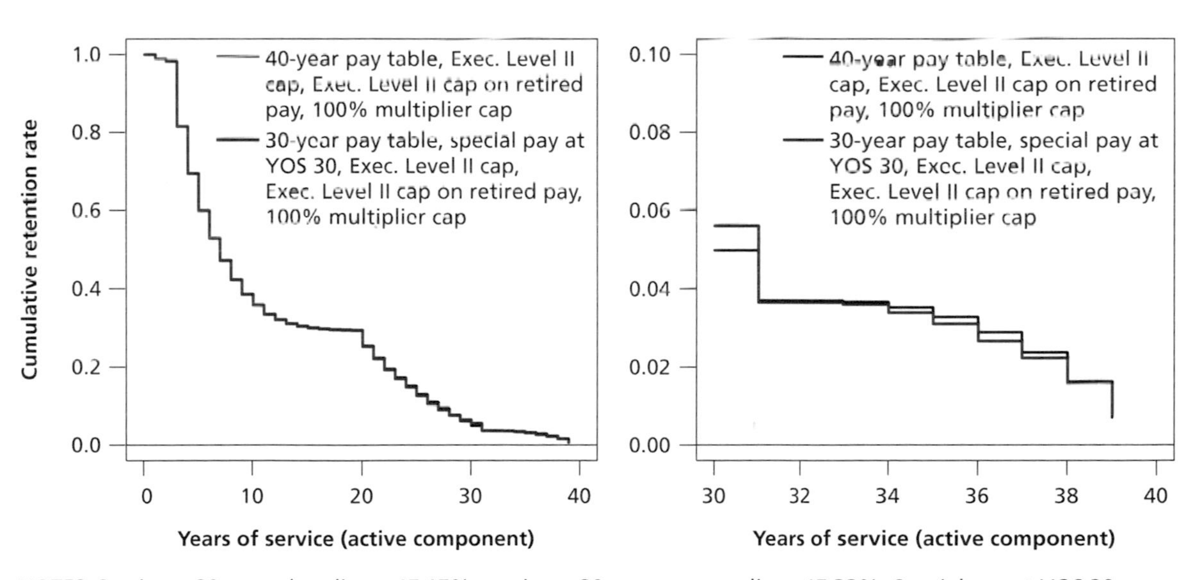

NOTES: Serving > 30 years, baseline = 17.17%; serving > 30 years, new policy = 17.33%. Special pay at YOS 30 = $99,600.

RAND *RR1209-6.6*

Figure 6.7
Simulated Effects on Retention of Reverting to the 30-Year Table and Adding Special Pay at YOS 30, Army Enlisted Personnel

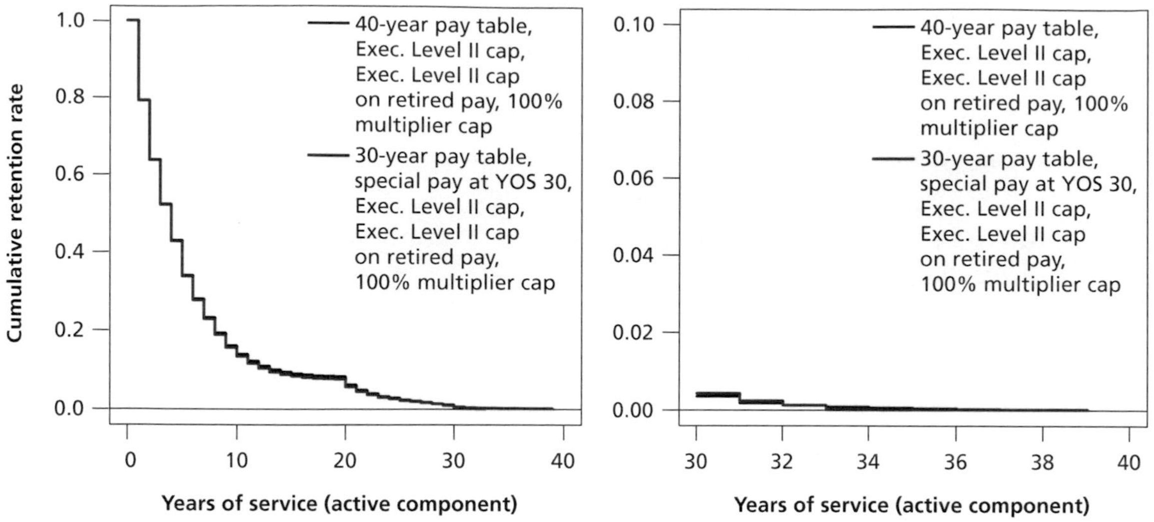

NOTES: Serving > 30 years, baseline = 2.66%; serving > 30 years, new policy = 2.89%. Special pay at YOS 30 = $37,400.

RAND *RR1209-6.7*

cial pay improves retention after YOS 30 slightly, but retention prior to YOS 20 drops slightly. It is likely that a more complex set of special pays, such as the addition of another gate pay between YOS 20 and YOS 30, would be needed to sustain retention across the force.[6]

Cost and Cost Savings Under a 40-Year Versus 30-Year Pay Table

The SASC report also requested information on the cost to DoD of operating under the 40-year pay table rather than a 30-year table adjusted for inflation. Answering this question requires the ability to compute costs had DoD operated under a 30-year table, all other factors held constant. The DRM simulations provide a method for making such a computation. Here, we provide the aggregate results for the active component for officers and enlisted personnel.

We use the DRM to compute current compensation costs, retired pay costs, and the total of current and retired pay costs under the 40-year table for each service and then aggregate the

[6] The estimate for the gate pay required to restore retention of enlisted personnel is conditioned on the estimates for the parameters of the DRM structural model. These estimates result in a higher implied discount rate for enlisted personnel than for officers, and also a higher implied posterior taste distribution for enlisted soldiers. Both of these may work to decrease the amount of gate pay needed to restore enlisted retention relative to officers.

A limitation of the current version of the DRM is that it uses a single point estimate for the discount rate, rather than allowing for individual heterogeneity in the discount rate over the enlisted and officer populations. A model with heterogeneity in both taste and discount rate would result in the YOS 20+ population being selected on both characteristics, which would result both in a higher posterior mean for taste and a higher posterior mean for the discount factor (that is, lower posterior mean for the discount rate). Such a model might well result in a higher estimate for the gate pay.

Thus, the estimates presented for gate pays should be regarded as good working estimates, subject to refinement given experience in retention behavior under a policy where both the 30-year pay table and a special pay to restore retention are implemented.

results. We then make similar computations under the 30-year table without sustaining the experience mix (as shown in Figures 6.4 and 6.5 for the Army) and again with the special and incentive pay to sustain the experience mix (as shown in Figures 6.6 and 6.7). Current compensation includes regular military compensation,[7] plus any relevant special pay. Retired pay is the accrual cost to DoD and the Department of the Treasury of nondisability retirement for DoD personnel. Note that in deriving these estimates, (1) current compensation costs are based on a simulated steady-state retention profile, which may differ from the actual realized retention profile in recent years, and (2) deferred compensation costs are based on an accrual charge that is calculated within the DRM, based on the simulated steady-state retention profile. All official computations of the accrual charge are performed by the DoD Office of the Actuary using a retention profile derived from average continuation rates in recent years. Thus, our estimates are not intended to replace official estimates for budget planning purposes. Instead, they are intended to inform the policy discussion. Should policymakers choose to change the pay table, DoD would and should provide estimates of how the costs would change. In addition, DoD may be asked to consider scenarios not covered in our analysis, such as how the costs would be affected by the currently planned force downsizing or by a decision to decrease and limit the number (or percentage) of officers and enlisted personnel with more than 30 years of service.

Table 6.1 shows that in the absence of a special pay to sustain the retention profile, enlisted costs per member for active-component personnel fall by 2.3 percent, while costs per member

Table 6.1
Costs and Cost Savings for the Active Component Under a 40-Year Versus 30-Year Pay Table

	40-Year Table	30-Year Table Without Sustaining Experience Mix		30-Year Table Plus Special Pay	
	Costs (2015 billions $)	Costs (2015 billions $)	% Change in Cost per Member	Costs (2015 billions $)	% Change in Cost per Member
Officer					
Current compensation	20.30	19.99	−1.5	20.23	−0.3
Retired pay	5.35	5.15	−3.6	5.24	−2.0
Total	25.64	25.15	−1.9	25.47	−0.7
Enlisted					
Current compensation	67.37	66.35	−1.5	66.87	−0.7
Retired pay	10.96	10.18	−7.2	10.47	−4.5
Total	78.34	76.53	−2.3	77.34	−1.3
Total					
Current compensation	87.67	86.35	−1.5	87.10	−0.6
Retired pay	16.31	15.33	−6.0	15.71	−3.7
Total	103.98	101.68	−2.2	102.81	−1.1

NOTE: Some costs may not add exactly due to rounding.

[7] Regular military compensation includes basic pay, the basic allowance for subsistence, the basic allowance for housing, and the federal tax advantage from receiving allowances tax-free. Regular military compensation is generally considered the military counterpart of a civilian salary.

for officers fall by 1.9 percent, for an overall decrease in costs of 2.2 percent, or $2.3 billion. Much of this savings is due to the reduction in the years of experience of retained personnel, as seen in Figures 6.4 and 6.5 for the Army.

The special pay increases current compensation costs, but current compensation is still slightly less costly under the 30-year table with special pay than under the 40-year table. For enlisted personnel, current costs per member are 0.7 percent lower, as seen in the last column of the table, but are only 0.3 percent lower for officers, for an overall decrease 0.6 percent for both officers and enlisted personnel. Part of this decrease is due to the slightly lower retention of enlisted personnel with fewer than 20 years of service. If a more complex set of special pays were used, such as the addition of another gate pay, retention would improve for this group, but the cost savings for enlisted personnel would of course be less than what we estimate. Because the special pay is not included in basic pay for the purposes of computing retired pay, retired pay costs decrease by 3.7 percent for the entire active component. Overall, costs are lower by 1.1 percent, or by about $1.2 billion. Again, if another special pay were added for enlisted personnel, the decrease would be less, though the complexity of the payment scheme would increase.

CHAPTER SEVEN

Conclusions

The qualitative and quantitative research, together with the review of the theoretical literature, allows us to draw several conclusions about the performance of the 40-year versus 30-year pay table in retaining senior personnel. We summarize those conclusions here.

Performance of the 40-Year Pay Table and Desirability of Reverting to the 30-Year Table

Our interviews indicated that the services have been able to retain adequate numbers of experienced personnel under the 40-year pay table, and similarly, they were able to do so under its predecessor, the 30-year pay table. Both the 40-year and 30-year pay tables have proven satisfactory overall in providing the services with the retention profiles needed by years of service, as well as with pools of personnel of sufficient quality and size from which to select senior leaders, though none of the interviewees provided specific quantitative evidence of how manpower utilization or productivity improved. Overall, none of the experts we spoke with felt that the 40-year pay table was necessary to successfully retain the most-experienced personnel, as long as none of the other 2007 changes in compensation is reversed.

But nearly all of the experts agreed that it is not desirable to return to the 30-year table, for a variety of reasons. Perhaps the most common reason given was the adverse effect on the morale of both senior personnel and more-junior personnel coming up the ranks, as well as their spouses. This would come on top of the effects of other recent changes in military compensation, such as the recent change to the military retirement system. In other words, service members would perceive a move back to a 30-year table as a cut in compensation at a time when other changes in compensation have occurred or are under discussion.

Another reason that interviewees gave for supporting the 40-year table was the need for flexibility to offer more-generous longevity increases in the most-senior grades, even if there is little need for that flexibility today or in the recent past. The argument here is that the lack of lateral entry into the military and the difficulty of replacing the most-senior personnel makes such flexibility important. And even though more-junior personnel could be promoted more quickly to fill these positions, interviewees preferred to retain the experienced people, especially because the services have other tools, such as mandatory separation rules, to induce these people to leave if the additional retention was not desirable. Others mentioned the increased requirements for senior personnel, not only as a result of operations in Afghanistan and Iraq but also because of the perception that the nature of military service is changing, now emphasizing technical skills and experience. Consequently, virtually all of the experts supported keeping the 40-year pay table.

Some interviewees indicated that reverting to a 30-year pay table could have adverse effects on retention and, therefore, special and incentive pay (such as assignment incentive pay) should be available to improve retention. Analysis of retention and cost using the DRM allowed us to ascertain whether retention could be sustained under a 30-year pay table relative to the 40-year pay table, how large of a special pay would be required to sustain retention, and at what cost. We conducted the analysis for each service but illustrated the results in the text for the Army only. We found that reverting to a 30-year pay table would adversely affect the retention of officers and enlisted personnel, especially among those with more than 30 years of service, but also among those with more than 20 years. We found that a special pay given at YOS 30 of between $87,900 and $99,600 for officers and between $37,400 and $58,700 for enlisted personnel (in 2015 dollars) would restore retention relative to the 40-year pay table, assuming no other change in compensation, at a cost savings for the active component of about $1.2 billion. Thus, if the 30-year table were brought back, a special pay that could be targeted to senior personnel would indeed be needed and would be effective.

Trends in the Use of Personnel with More Than 30 Years of Service

Our review of the historical context of the 2007 move to the 40-year pay table indicated that the main impetus for change was to permit longer careers and increased retention among general and flag officers. Several interviewees indicated that then-Secretary Rumsfeld believed that the 30-year pay table was an obstacle to these goals. We analyzed retention trends before and after 2007 and found that the number of senior personnel with more than 30 years of service did in fact increase, but in percentage terms, the greatest increase was not among GOFO personnel but among senior enlisted personnel and field grade officers. This does not mean that the move to the 40-year pay table did not better facilitate longer careers and increase GOFO retention. In fact, analysis of trend data does not permit any causal interpretation, so by considering only the trend data alone, it is unclear whether personnel management improved. Rather, the trend data reveal that much of the increase in retention occurred among populations that were not the explicit target groups of the legislation. On the other hand, the number of the most-senior officers did increase, consistent with the legislation.

In 2007, 4,175 active-duty service members had more than 30 years of service. This figure rose by 58 percent, to 6,583, by 2014. Most of this increase is attributable to increases in enlisted personnel (from 809 to 2,029), especially E-9s, and officers in the grades O-4 to O-6 (from 2,242 to 3,354), especially O-5s. While we did not investigate how many of these officers have prior enlisted service, it is likely that many do, because O-6s are not permitted to serve beyond 30 years of commissioned service without a waiver. The number of senior officers in grades O-7 to O-10 also increased, but by a far more modest amount in percentage terms (from 583 to 621 over this period); in fact, the number of senior officers declined between 2013 and 2014.

Furthermore, we find that the increase in the number of personnel with more than 30 years of service began before 2007, especially for enlisted personnel, suggesting that the increases after 2007 were part of an ongoing trend. For example, the number of enlisted personnel began to increase in 2006, following a period of decline between 2002 and 2005. The increase in officer strength did begin in 2007, but that trend masked some differences across grades. For example, the number of O-4s with more than 30 years of service had been increasing since 2003. Similarly, the increase in the number of warrant officers also began in 2007, but that increase masked tremendous differences across grades. The number of W-5s actually decreased between 2007 and 2011, but that decrease was more than offset by increases in the number of W-4s over that period.

Interviewees mentioned that, as a result of increased requirements driven by military operations in the Middle East since September 11, 2001, there was an increased demand for personnel with more than 30 years of service. These requirements increased the need for personnel with strong leadership skills and experience, technical knowledge, and the ability to command the battlefield and effectively support those operations. The trends in the number of personnel with more than 30 years of service show that the services clearly made more use of senior personnel to meet these requirements, including not just senior officers but also field grade officers in O-4 to O-6, warrant officers, and senior enlisted personnel.

We also considered five-year continuation rates to YOS 30 among those with 26 years of service, as well as two-year continuation rates to YOS 32 among those with 30 years of service. We found considerable variation in continuation rates, but little evidence of a marked increase in rates after 2007. The lack of a marked change in continuation rates after 2007, despite the increase in the numbers of personnel with more than 30 years of service, suggests that the increase in the number of senior personnel may be at least in part a function of an increase in the overall size of the military since 2001, driven by increased demands and pace of deployment in this period, rather than a true increase in the percentage of personnel choosing to stay in the military past 30 years. If true, this would be consistent with the perception of our interviewees that the legislative changes in 2007 had little effect on the retention decisions of the majority of service members.

Insights from the Theoretical Literature

We also find that the theoretical literature supports a skewed compensation structure that provides incentives to stay, especially among the most-talented individuals, and to continue to perform and supply effort through YOS 40. That is, the literature provides little support for ending those incentives at YOS 30. This is especially important if the increased demand for personnel with more than 30 years of service is expected to continue in the future. While much of the skewness of the current compensation system occurs through the military retirement system for the most-senior officers (in grades O-9 and O-10) rather than through basic pay increases, given that basic pay increases are not realized for these personnel as a result of the Executive Level II pay cap, basic pay increases also increase retirement benefits. Furthermore, basic pay increases in the pay table do affect skewness for senior enlisted personnel. That said, the literature does not indicate how much skewness is adequate. This issue was also raised by some of the interviewees who questioned how much pay is enough to meet retention needs, and how much retention of the most-senior personnel is enough to meet requirements.

The Cap on Basic Pay for Computing Retired Pay

Finally, while the focus of our project was the pay table, the other compensation changes that occurred in 2007 were a topic that came up over the course of our study and one we considered in our DRM analysis. As mentioned in the introduction, four changes occurred in 2007, including increasing the cap on basic pay, removing the cap for computing retired pay, and removing the cap on years of service for computing the retired pay multiplier. In the 2015 NDAA, the cap on basic pay for computing retired pay was restored for service after December 2014. This 2014 change garnered considerable comment during our interviews. Most considered it a more important issue than the pay table per se, and many expressed concern about the effect of this change on the retention of senior officers, with some interviewees being adamant that the change should be reversed before readiness was significantly affected. Others were more uncertain about how retention would be affected and whether any change in retention, if such a changed occurred, was something to worry about, noting that it depended on the services' requirements for these personnel and the value of additional experience in the military. We simulated the effects of the 2014 change on officer retention and found that restoring the cap, as was the case in 2014, had virtually no effect on the retention of Army officers with fewer than 34 years of service in the steady state. However, retention fell among those with 34 or more years of service, and the number of officers with more than 30 years of service as a percentage of those with more than 20 years decreased from 16.4 percent to 14.6 percent. Thus, the DRM results show a drop in retention among the most-senior officers. And because the cap has no effect on enlisted personnel, we find no effect and do not show the results for enlisted personnel.

Closing Thoughts

In sum, reverting to a 30-year table for senior personnel without an additional special pay would hurt the retention of senior leaders. While both a 40-year and 30-year table could be equally effective in sustaining retention (as long as the services would have adequate special pay to manage retention under a 30-year table), continuing with the 40-year table is preferred. It performs well, and many argue that it improves readiness and flexible personnel management; in contrast, many felt that reverting to the 30-year table could adversely affect morale and perceptions about the stability and value of military compensation overall. Furthermore, given the increase in the number of personnel with more than 30 years of service since 2007, reverting to a 30-year table would affect far more people—58 percent more people—than it did when a 30-year table was in effect prior to 2007. Finally, the cost of maintaining the 40-year table is relatively small. Using the DRM, our estimate of the change in cost of keeping the 40-year table versus sustaining retention under a 30-year table with the use of special pay is about $1.2 billion for the active component, or a change of 1.1 percent in active-component personnel costs. (As mentioned, DoD would be the source of final cost estimates for budget planning purposes.) Given the disadvantages of reverting to a 30-year table and the advantages of keeping the 40-year table, these considerations suggest that continuing with the 40-year pay table is a sensible course of action.

APPENDIX

Tabulations by Service and Pay Grade

Chapter Five presented trends in the number of active-duty personnel with more than 20 and 30 years of service, by service. Here, we provide more-detailed tabulations by service and pay grade. We discuss each service separately. In addition, Chapter Five presented trends in continuation rates both to YOS 30 among those with 26 years of service and to YOS 32 among those with 30 years of service. This appendix presents tabulations of these rates by service.

Tabulations of Number of Personnel with More Than 30 Years of Service

Army

Figures A.1 and A.2 show the number of O-4 to O-10 Army officers with more than 30 years of service over the FY 2000 to FY 2014 period. As was the case for the overall trends, the largest increase in officers with more than 30 years of service was among the O-6s and O-5s, with a smaller increase among O-7s and O-8s. Figure A.1 shows a sizable increase in the number

Figure A.1
Number of Army Personnel with More Than 30 Years of Service, O-4 to O-6

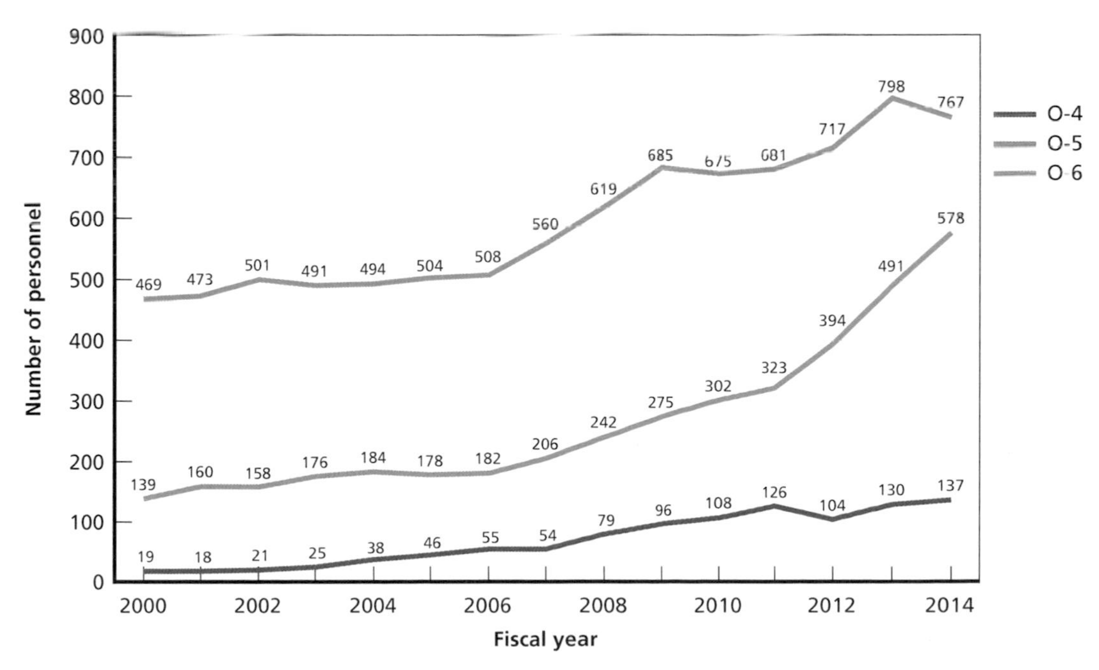

RAND RR1209-A.1

Figure A.2
Number of Army Personnel with More Than 30 Years of Service, O-7 to O-10

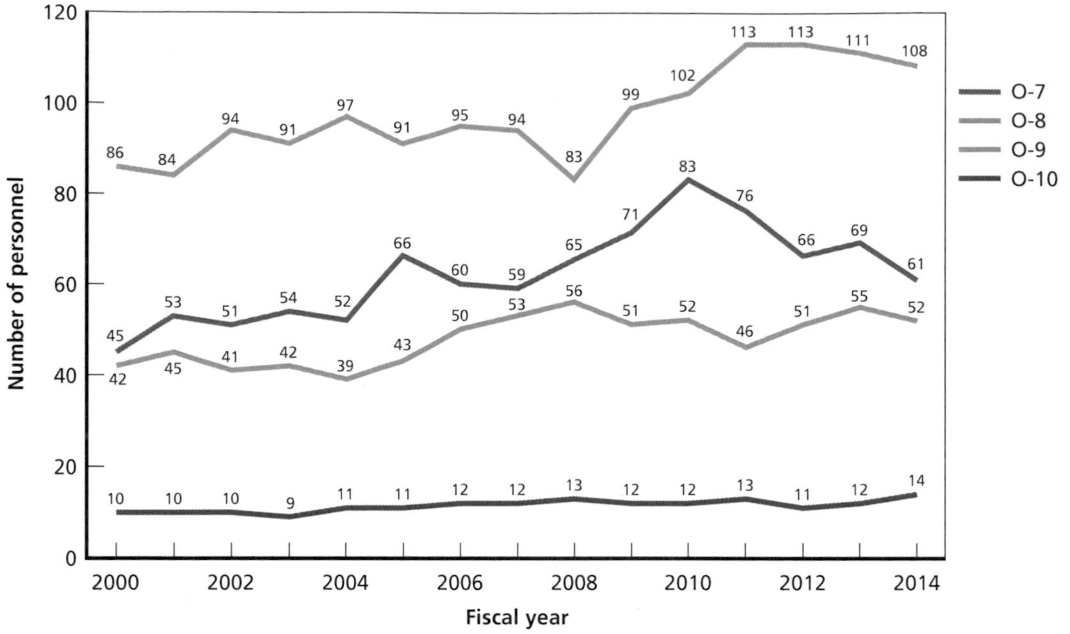

RAND *RR1209-A.2*

of O-6s with more than 30 years of service that begins in FY 2007. The number of O-6s was about 500 for several years prior to 2007. Starting in FY 2007, this number rose to 560, peaking at 798 in FY 2013. Similarly, the number of O-5s was at about 180 prior to FY 2007, reaching 578 by FY 2014. As noted, these increases may represent prior enlisted personnel who had greater incentive to stay as a result of contextual factors, including the weak economy and the wars in Afghanistan and Iraq (as well as legislative changes), or they could possibly be recalled retirees who have more than 30 years of service.

Changes in the number of officers in grades O-7 to O-10 with more than 30 years of service changed more modestly, if at all (Figure A.2). The number of O-8s was at 95 in FY 2006 and 83 in FY 2008, then rose to 113 in FY 2011 before falling slightly to 108 in FY 2014. The number of O-7s with more than 30 years of service rose from the low 50s in the early 2000s to 83 in FY 2010. However, this number has fallen steadily since, to 61 in FY 2014. Changes in the number of O-9s and O-10s with more than 30 years of service have been minor, and there is no markedly upward trend in personnel with more than 30 years of service for these pay grades.

Figure A.3 shows the number of W-3 to W-5 Army warrant officers with more than 30 years of service over the FY 2000 to FY 2014 period. There has been no change for W-3s. There were 145 W-4s in FY 2002, and this number then fell over the next four years to reach 71 in FY 2006. The number of W-4s with more than 30 years of service then rose after 2007, to 101 in FY 2010, before falling again to 62 in FY 2014, well below its 2002 level. The number of W-5s has increased since 2007, but this increase is part of a longer upward trend, and the majority of the increase occurred after FY 2012. The number of W-5s rose from 198 in FY 2006 to 218 in FY 2007. It fell slightly afterward, was still at 218 in FY 2012, and then rose to 273 in FY 2014. In this case, the most significant increase in the number of W-5s with 30 years of service does not seem to be a result of the 2007 changes to the pay tables.

Figure A.3
Number of Army Personnel with More Than 30 Years of Service, W-3 to W-5

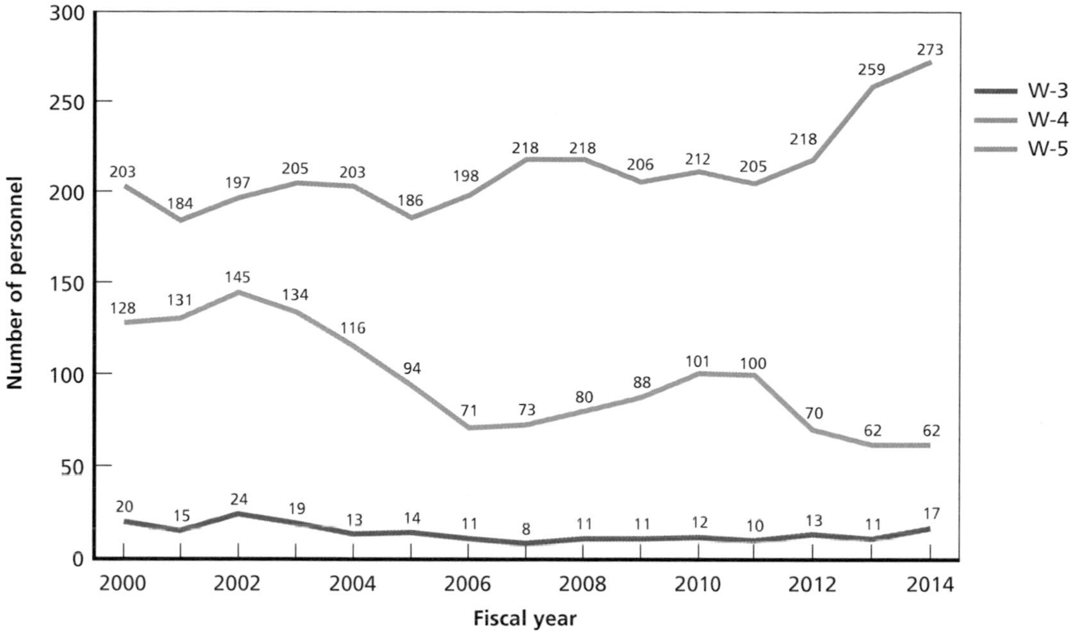

RAND *RR1209-A.3*

Finally, Figure A.4 shows the number of E-5 to E-9 Army enlisted personnel with more than 30 years of service over the FY 2000 to FY 2014 period. The figure confirms that the majority of the increase in personnel with more than 30 years of service was driven by E-9s, although there are also upward trends for E-8s, E-7s, and E-6s (some of these may be coding errors). Importantly, in each case, the upward trends begin prior to FY 2007. Focusing on the upward trend in E-9s, the figure shows a rather sharp increase starting after about FY 2009. The number of E-9s with more than 30 years of service rose from 143 in FY 2001 to 789 by FY 2014.

Air Force

Figures A.5 and A.6 show the number of O-4 to O-10 Air Force officers with more than 30 years of service over the FY 2000 to FY 2014 period. Unlike in the Army, we do not see an increase in the number of O-6s; however, there is an increase in O-5s and O-4s. The number of O-6s with more than 30 years of service decreases steadily after FY 2003. The number of O-5s also fell from FY 2002 to FY 2007. However, it then rose from 128 in FY 2007 to 197 in FY 2014. The number of O-4s rose in the early 2000s but then fell from FY 2003 to FY 2007. It rose again, more steadily, from 11 in FY 2007 to 93 in FY 2014. For senior officers, only the O-9s show a steady increase after 2007. The number of O-7s with more than 30 years of service decreased steadily after FY 2005, from 39 in FY 2005 to 19 by FY 2014. The number of O-8s also declined over the period under consideration. This number was at 68 in FY 2005 and peaked at 91 in FY 2009. It then fell steadily to 48 by FY 2014. The number of O-9s with more than 30 years of service did increase after 2007, but only slightly. This number was at 39 in FY 2005 and then rose to 46 by FY 2014. The number of O-10s was small and remained steady over this period.

Figure A.4
Number of Army Personnel with More Than 30 Years of Service, E-5 to E-9

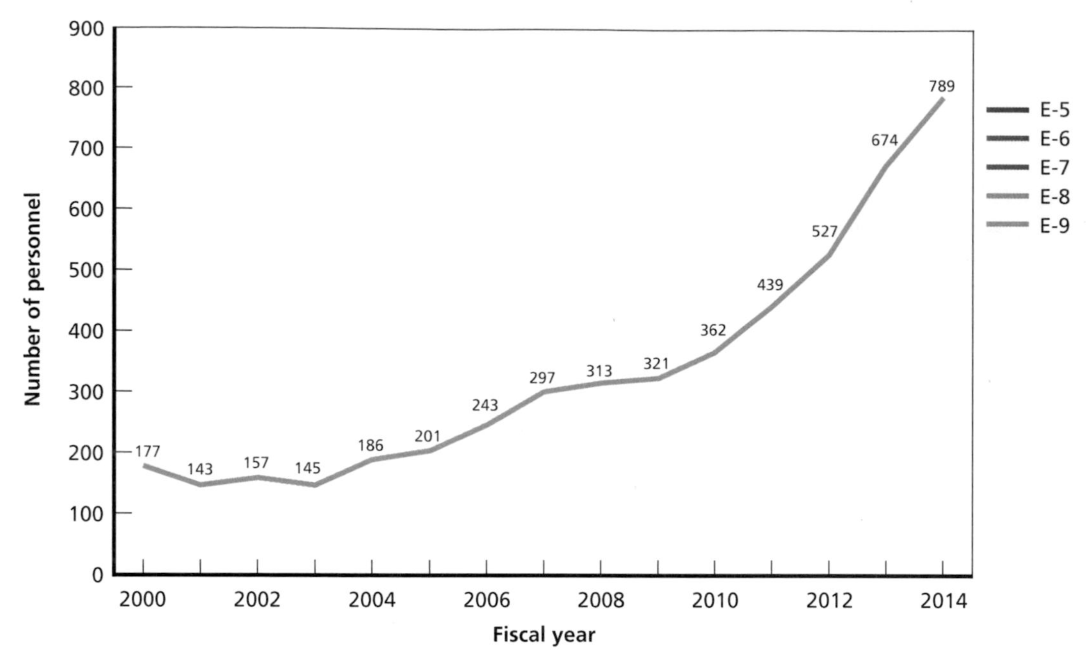

RAND *RR1209-A.4*

Figure A.5
Number of Air Force Personnel with More Than 30 Years of Service, O-4 to O-6

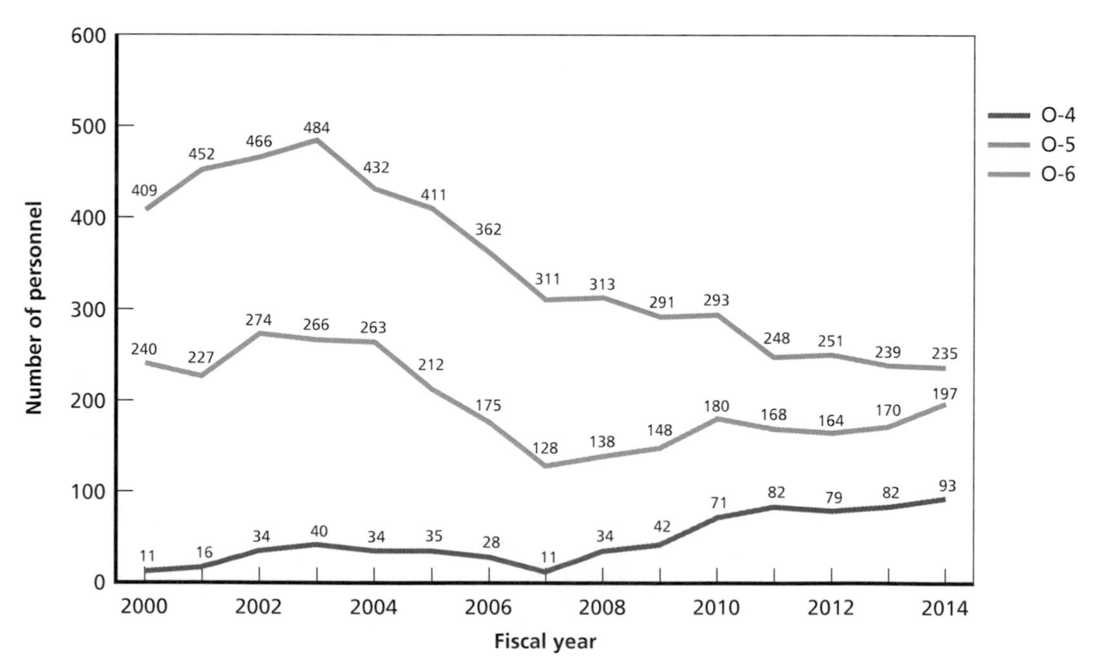

RAND *RR1209-A.5*

Figure A.6
Number of Air Force Personnel with More Than 30 Years of Service, O-7 to O-10

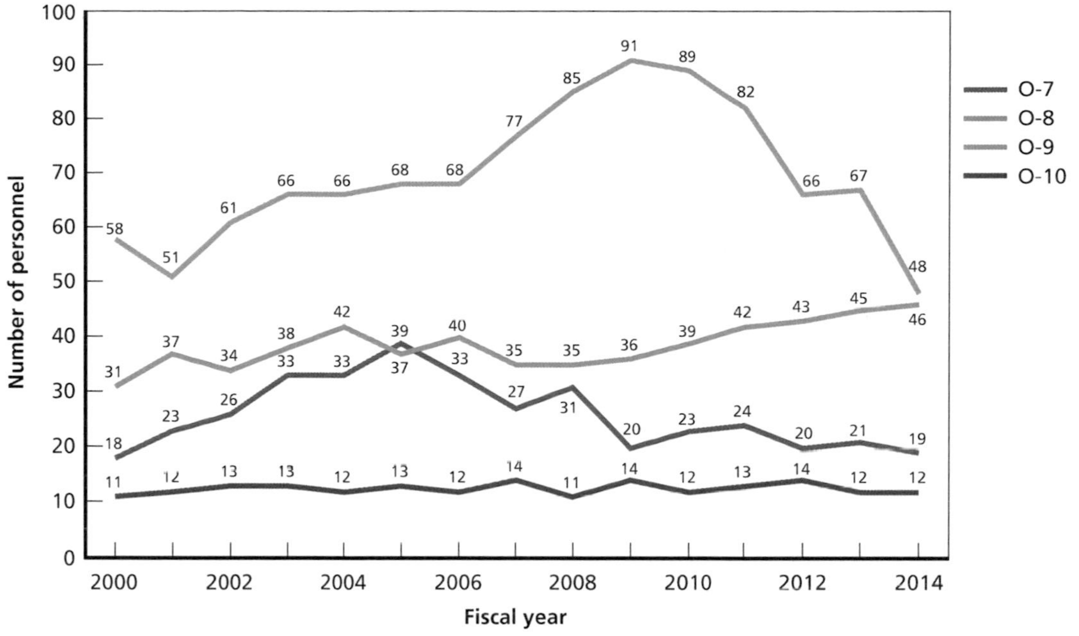

RAND *RR1209-A.6*

Figure A.7 shows the number of E-5 to E-9 Air Force enlisted personnel with more than 30 years of service. The figure shows that the increase in the number of enlisted personnel was driven almost entirely by an increase in E-9s. However, this increase did not really begin until FY 2008. The number of E-9s with more than 30 years of service was at 117 in FY 2005. It rose to 143 in FY 2007 but fell to 113 in FY 2008. The number then rose to 234 by FY 2014. The increase could be a delayed response to legislative changes in 2007 or a result of other contextual factors, or a combination of both.

Marine Corps

Figures A.8 and A.9 show the number of O-4 to O-10 Marine Corps officers with more than 30 years of service over the FY 2002 to FY 2014 period. Figure A.8 shows a small post-2007 increase in O-6s, but no real change for O-4s or O-5s. The number of O-6s with more than 30 years of service fell from FY 2003 to FY 2008 (from 138 to 92). It then rose to 112 in FY 2014, which was still below its FY 2003 level. Looking at senior officers (Figure A.9), the number of O-7s decreased from 36 in FY 2005 to 15 in FY 2014. The number of O-8s with more than 30 years of service increased between FY 2005 and FY 2014, though by only ten personnel. There was little or no change in the number of O-9s and O-10s.

Figure A.10 shows the number of W-3 to W-5 Marine Corps warrant officers with more than 30 years of service over the FY 2002 to FY 2014 period. The figure shows an increase in both the number of W-5s and the number of W-4s. The number of W-5s with more than 30 years of service fell between FY 2005 and FY 2010 from 16 to five. However, this number has risen steadily since then, to 24 in FY 2014. The number of W-4s also increased starting in FY 2010, from two in FY 2009 to 16 in FY 2014. Again, while this is an increase, it is a very small number of personnel. It is also an increase that does not immediately follow the 2007 legislative change.

Figure A.7
Number of Air Force Personnel with More Than 30 Years of Service, E-5 to E-9

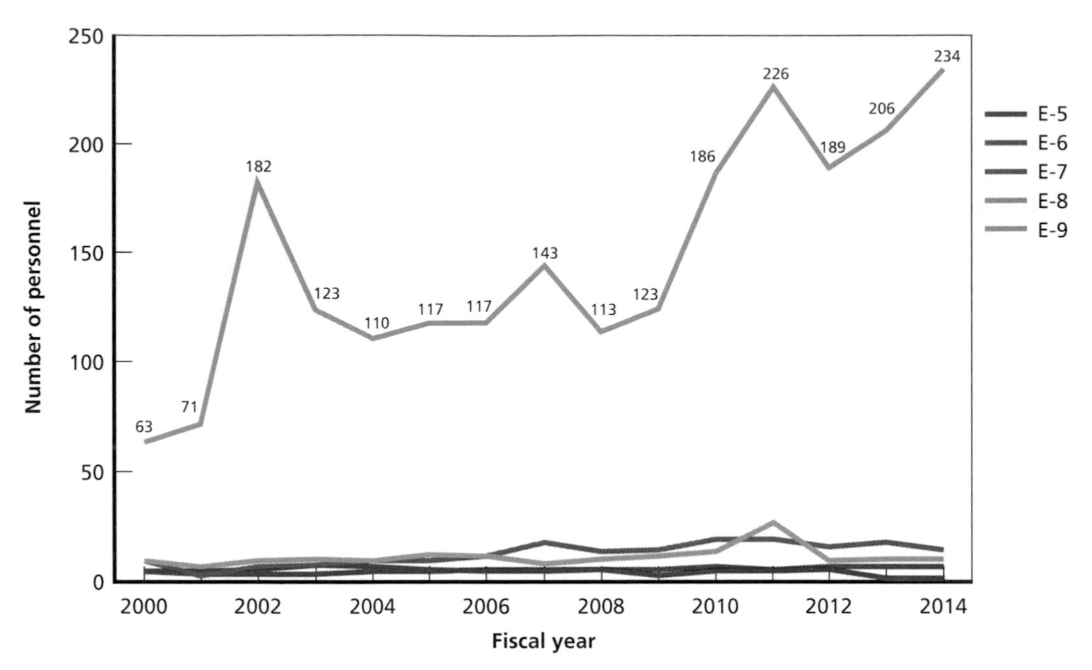

RAND *RR1209-A.7*

Figure A.8
Number of Marine Corps Personnel with More Than 30 Years of Service, O-4 to O-6

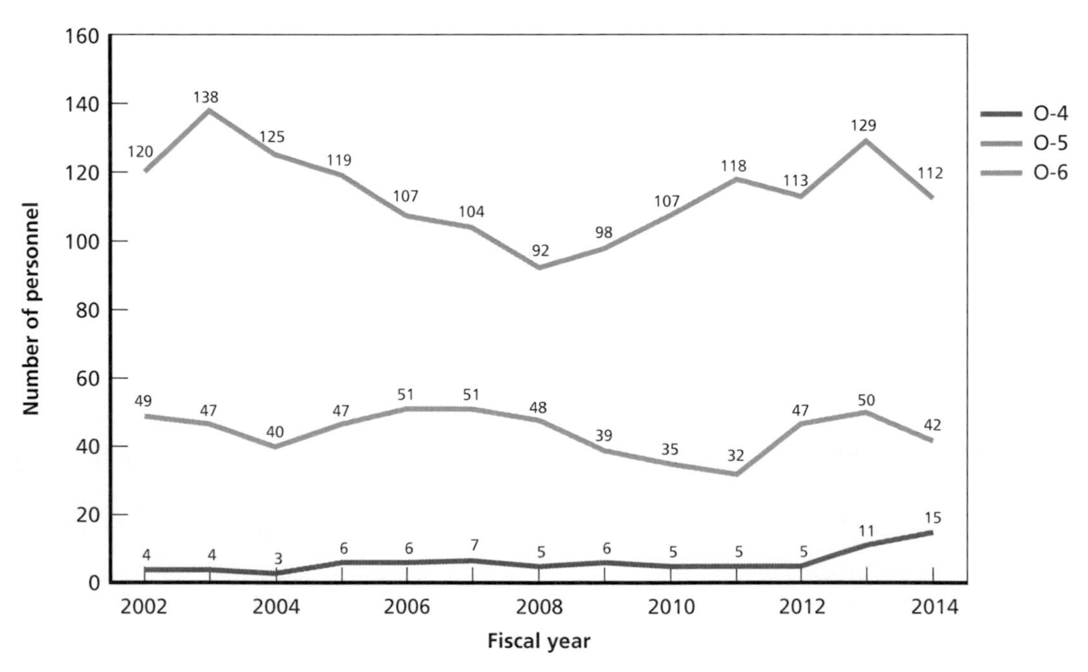

RAND *RR1209-A.8*

Figure A.9
Number of Marine Corps Personnel with More Than 30 Years of Service, O-7 to O-10

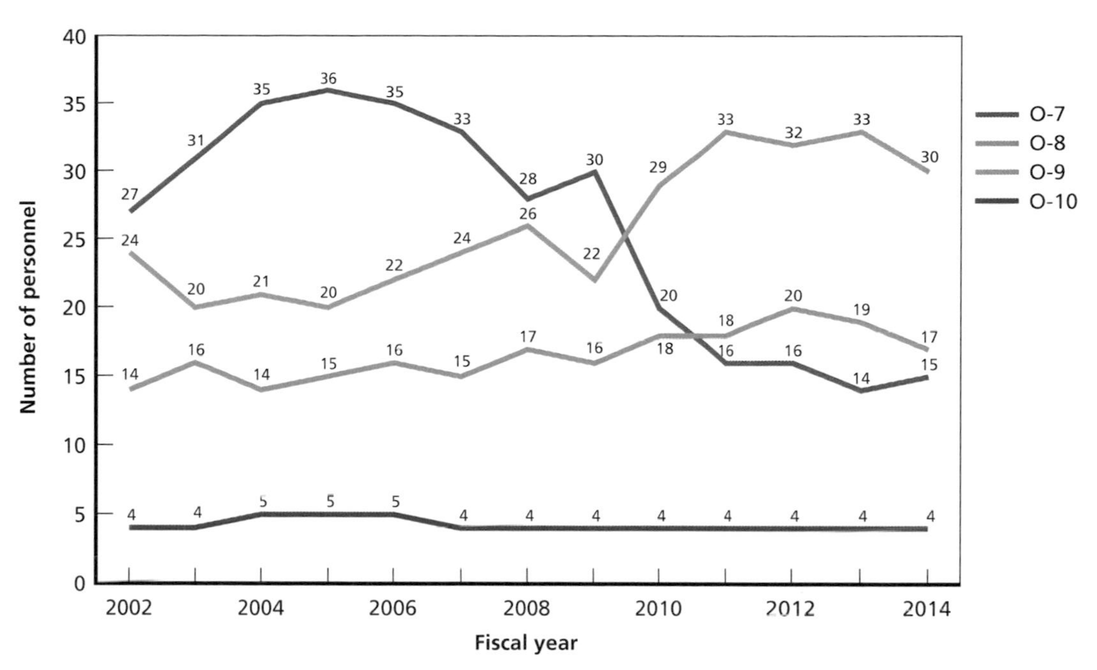

RAND *RR1209-A.9*

Figure A.10
Number of Marine Corps Personnel with More Than 30 Years of Service, W-3 to W-5

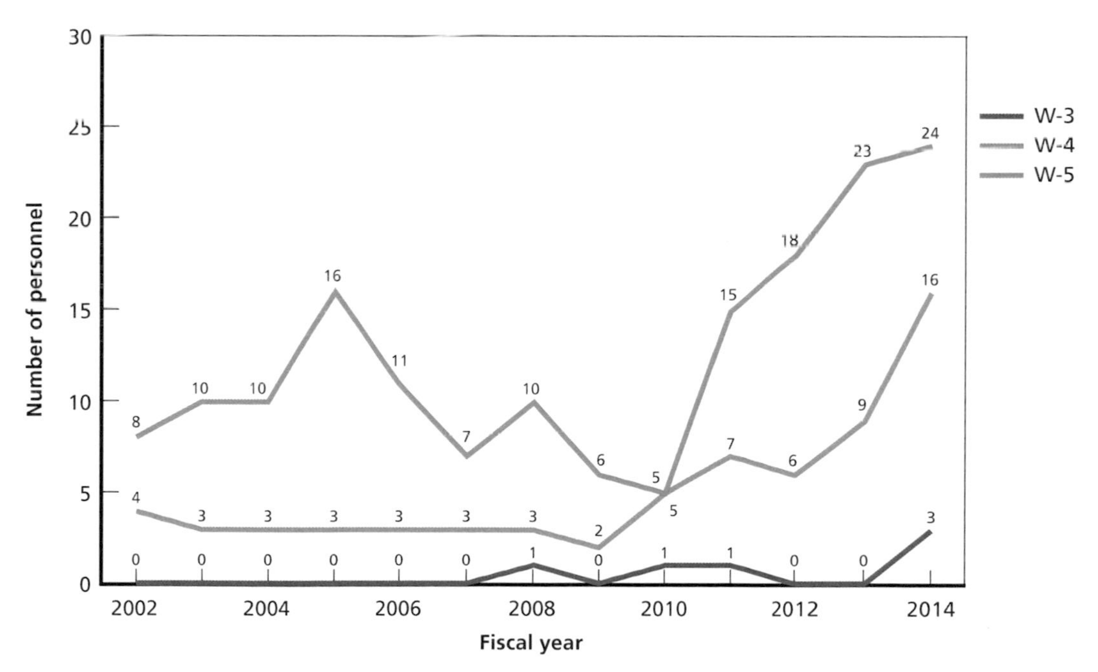

RAND *RR1209-A.10*

Finally, Figure A.11 shows the number of E-5 to E-9 Marine Corps enlisted personnel with more than 30 years of service. The figure confirms that the increase occurred almost entirely among E-9 personnel. However, the number of E-9s increases steadily from FY 2004 on, starting at 46 in FY 2004 and peaking at 166 in FY 2013. According to Marine Corps representatives we interviewed, many of these E-9s were asked to stay to fill specific billets during conflicts in Afghanistan and Iraq.

Navy

Figures A.12 and A.13 show the number of O-4 to O-10 Navy officers with more than 30 years of service over the FY 2000 to FY 2014 period. Figure A.12 shows little change in the number of O-6s, falling from 524 in FY 2000 to 466 in FY 2014. In contrast, the number of O-5s with more than 30 years of service rose steadily over this period, from 238 in FY 2001 to 522 in FY 2014. Finally, there is also an increase in O-4s in the Navy. This number rose from 31 in FY 2004 to 182 by FY 2014.

Looking at senior officers (Figure A.13), as in the other services, we see no real change in the number of O-10s with more than 30 years of service. There is also little change in the number of O-9s. This number hovered around 30 until FY 2007, rose to 45 in FY 2013, then fell to 36 in FY 2014. The number of O-8s with more than 30 years of service was at 59 in FY 2005 and 66 in FY 2014. Overall, this is again a small change. The number of O-7s does seem to have increased markedly since 2007. The number was at 35 in FY 2006 and 31 in FY 2007, but rose steadily afterward, to 81 in FY 2014.

Figure A.11
Number of Marine Corps Personnel with More Than 30 Years of Service, E-5 to E-9

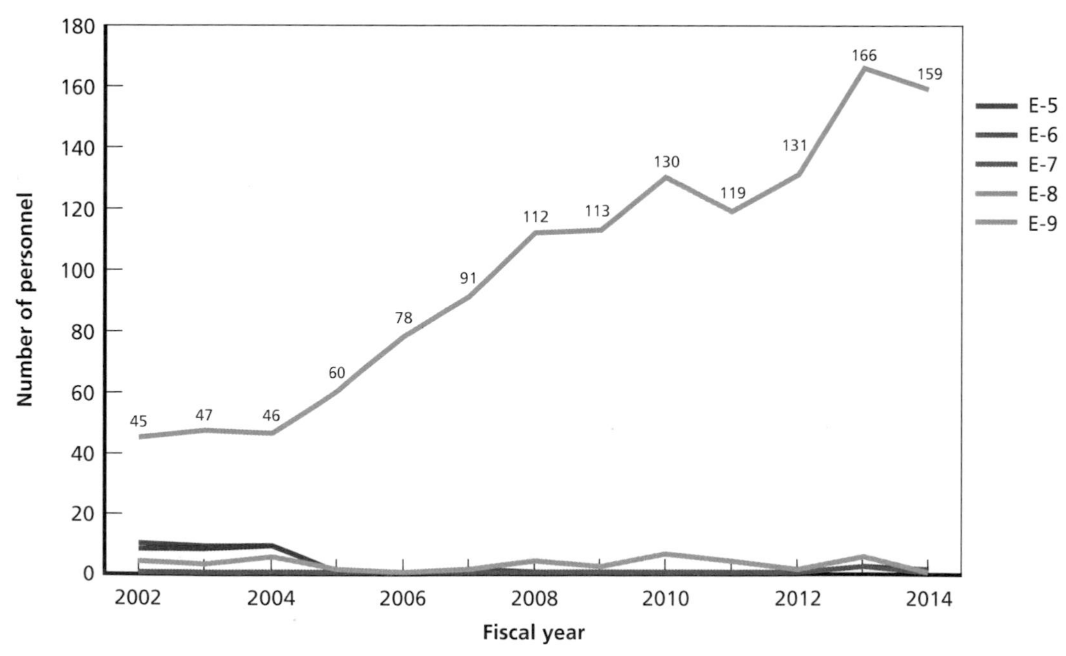

RAND *RR1209-A.11*

Figure A.12
Number of Navy Personnel with More Than 30 Years of Service, O-4 to O-6

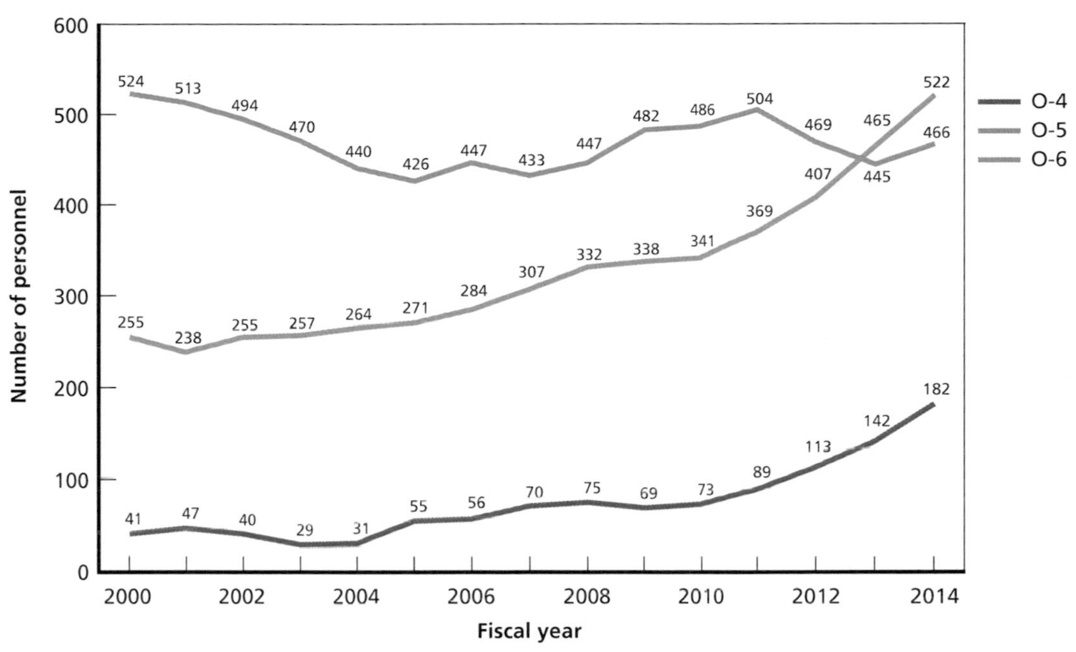

RAND RR1209-A.12

Figure A.13
Number of Navy Personnel with More Than 30 Years of Service, O-7 to O-10

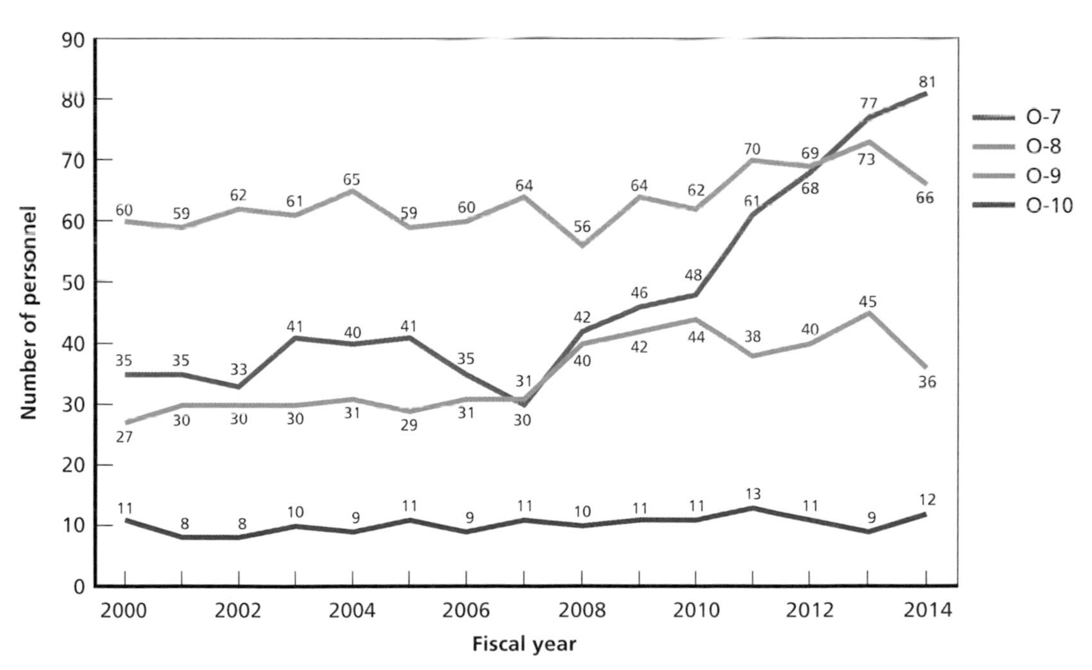

RAND RR1209-A.13

Figure A.14 shows the number of W-3 to W-5 Navy warrant officers with more than 30 years of service over the FY 2000 to FY 2014 period. The figure shows an increase on net for W-3s, W-4s, and W-5s. For W-3s, the increase seems to be part of a longer trend, rising from two in FY 2004 to 15 by FY 2008 and 19 by FY 2014. The number of W-4s with more than 30 years of service fell from 37 in FY 2006 to 21 in FY 2007, then rose to 58 in FY 2014. Finally, the number of W-5s was more or less steady between FY 2005 and FY 2009, although it had risen sharply between FY 2003 and FY 2005. However, the number fell from 40 to 21 between FY 2009 and FY 2011 and then rose again to 56 by FY 2014. Again, these changes do not seem to be directly linked to the legislative changes that occurred in 2007.

Finally, Figure A.15 shows the number of E-5 to E-9 enlisted personnel with more than 30 years of service over the FY 2000 to FY 2014 period. The figure shows that, as in other services, the increase in enlisted personnel with more than 30 years of service is driven largely by changes in the number of E-9s. While the number of E-9s with more than 30 years of service declined from FY 2000 to FY 2004, it increased from 129 to 205 by FY 2007. It then decreased to 161 by FY 2009 and then rose again to 237 in FY 2013. There were 204 of these E-9 personnel in the Navy in FY 2014.

Figure A.14
Number of Navy Personnel with More Than 30 Years of Service, W-3 to W-5

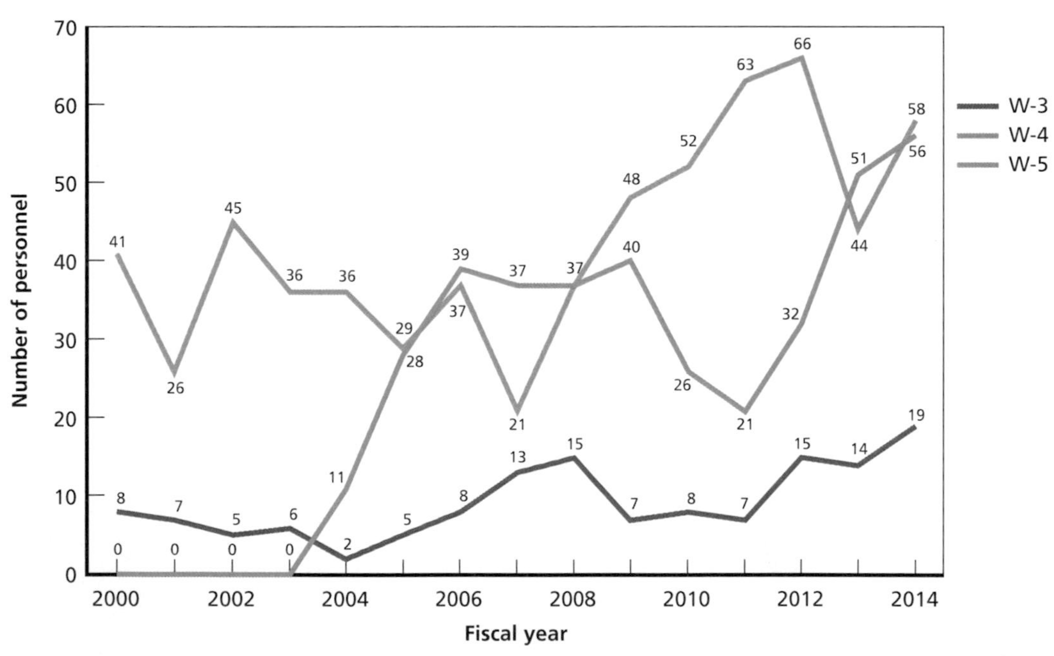

RAND *RR1209-A.14*

Figure A.15
Number of Navy Personnel with More Than 30 Years of Service, E-5 to E-9

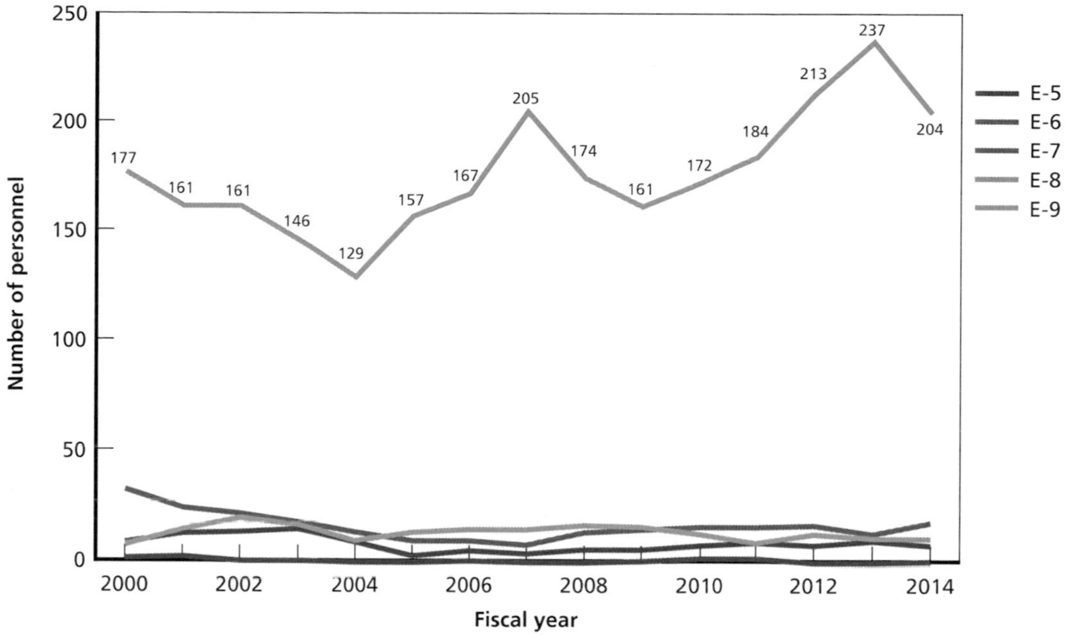

RAND RR1209-A.15

Analysis of Continuation Rates by Service

Army Continuation Rates

Figures A.16 through A.18 show continuation rates for Army officers, warrant officers, and enlisted personnel. For officers (Figure A.16), changes in continuation rates are less significant than for other types of personnel, particularly for more-senior personnel. Looking first at continuation rates to YOS 30 for personnel with 26 years of service, the rates increase slightly over the period under consideration, from about 41 percent before 2005 to between 45 and 53 percent from 2006 to 2010, and falling to 41 percent in 2011. Comparison of average continuation suggests some increase in the post-2007 period, from 43 percent prior to 2007 to 48 percent afterward. However, there is little evidence of an upward trend in this case, and even less evidence of an upward trend that is directly linked to the 2007 change in the pay table. For officers with 30 years of service, continuation rates to YOS 32 do not change significantly, remaining between 45 and 50 percent throughout the period under consideration. This lack of change is important, because increasing the retention of these senior officers was one of the main goals of the shift to the 40-year pay table.

Figure A.17 shows continuation rates for warrant officers in the Army. Again, there is little evidence of a significant upward trend in continuation rates for personnel at 26 or 30 years of service. For personnel with 26 years of service, continuation rates to YOS 30 remain between 40 and 50 percent over the period under consideration, and there is little evidence of a sustained trend in either direction, although the average continuation does appear to increase slightly after 2007, from 45 to 48 percent. For personnel with 30 years of service, continuation

Figure A.16
Continuation Rates, Army Officers

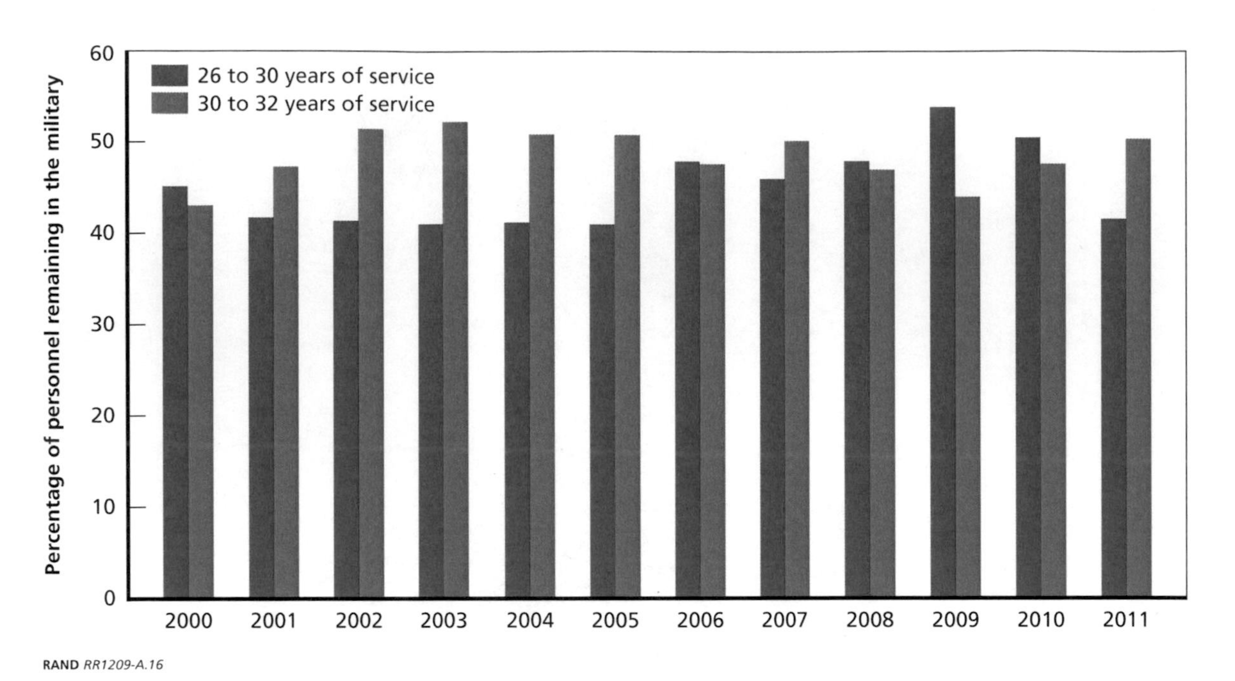

RAND *RR1209-A.16*

Figure A.17
Continuation Rates, Army Warrant Officers

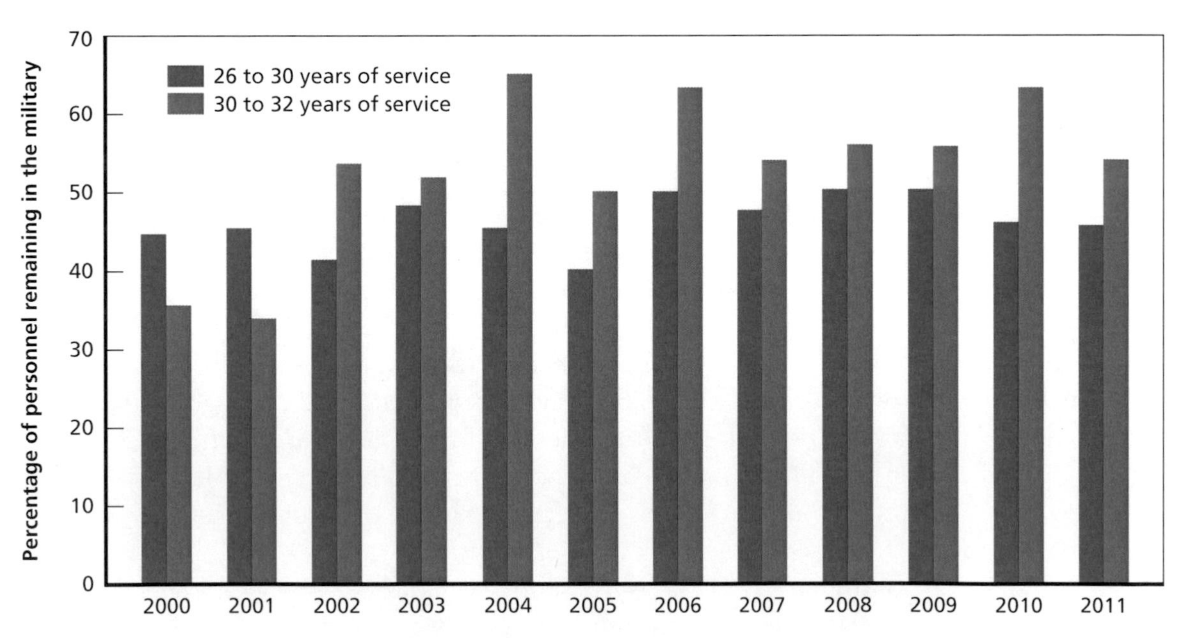

RAND *RR1209-A.17*

rates to YOS 32 do appear to increase between 2000 and 2011, but this increase occurs primarily in the period between 2000 and 2002. After 2002, continuation rates for these individuals vary between 50 and 65 percent, with no sustained trend and considerable variation from year to year. There is a larger increase in the post-2007 average continuation rate, from 50 percent for the pre-2007 period to 57 percent after 2007. This increase is driven largely by the lower continuation rates in 2000 and 2001.

For enlisted personnel, there had been some increase in continuation rates for people with 30 years of service to YOS 32 (Figure A.18). While continuation rates for these personnel remained between 22 percent and 46 percent between 2000 and 2005, a larger percentage of personnel remained in the service in years after 2006. Specifically, 55 percent of those with 30 years of service in 2006 remained in the Army through YOS 32, and 58 percent of those with 30 years of service in 2007 remained through YOS 32. This rate stayed between 54 and 58 percent for the remainder of the period under consideration. The average continuation rate to YOS 32 for personnel with 30 years of service increased from 37 percent prior to 2007 to 57 percent after 2007. However, it is worth noting that the increase in the continuation rate began in 2006 rather than with the legislative change in 2007. There was no real change in continuation rates for personnel with 26 years of service to YOS 30. In 2001, 31 percent of such personnel remained in the Army through YOS 30. This rate fell to 25 percent in 2005, peaked at 39 percent in 2009, and fell back to 31 percent in 2011. The pre- and post-2007 continuation rates also remained mostly the same for this group.

Overall for the Army, then, there is limited evidence that the change to a 40-year pay table substantially affected continuation rates for personnel at 26 or 30 years of service. Only enlisted personnel with 30 years of service showed any sustained upward trend in continuation rates over this period, and even this increase was relatively modest in size and not clearly linked to the legislative changes in 2007.

Figure A.18
Continuation Rates, Army Enlisted Personnel

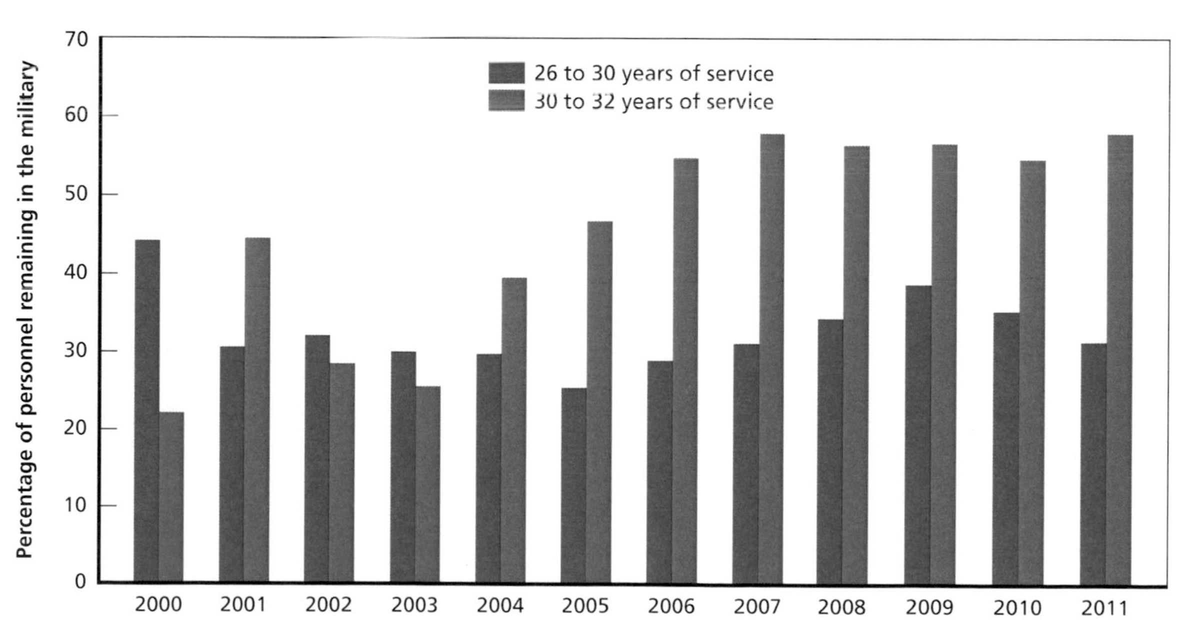

RAND *RR1209-A.18*

Air Force Continuation Rates

Figures A.19 and A.20 show continuation rates for Air Force officers and enlisted personnel. Figure A.19 shows continuation rates for officers. For officers with 26 years of service, continuation rates to YOS 30 appear to have fallen since 2000 and in the period after 2007. For officers with 26 years of service in years prior to 2002, continuation rates to YOS 30 stood at nearly 40 percent. However, this fell in 2003 to 33 percent and in 2006 to 31 percent. This continuation rate fell further after 2007, to 26 percent in 2011. A comparison of average continuation rates confirms the decrease, from 33 percent prior to 2007 to 28 percent afterward. For officers with 30 years of service, continuation rates to YOS 32 have shown no significant trend in either an upward or downward direction and have fluctuated somewhat over the time period under consideration. In the period prior to 2007, this rate varied between 33 and 37 percent (with the exception of 2004, when it was at 45 percent). After 2007, it varied between 29 and 41 percent. The average continuation rates also did not change much before and after 2007.

For enlisted personnel with 26 years of service, continuation rates to YOS 30 appear to decrease over the time period under consideration (Figure A.20). While 25 percent of Air Force enlisted personnel with 26 years of service in years prior to 2003 remained in the service to YOS 30, this continuation rate fell to 20 percent in 2004 and 2005 and then remained below 20 percent through 2011. For personnel with 30 years of service, continuation rates to YOS 32 have also not changed substantially. This rate has fluctuated between 3 and 6 percent over most of this period, with a few exceptions, including 2000 (7 percent), 2003 (13 percent), and 2006 (7 percent). However, these appear to be random fluctuations rather than any significant trend.

As in the case of the Army, there is no evidence of an increase in continuation rates for personnel with 26 or 30 years of service in the period under consideration. In fact, there is even evidence that continuation rates have fallen for certain groups of senior personnel, including continuation to YOS 30 of officers and enlisted personnel with 26 years in the military.

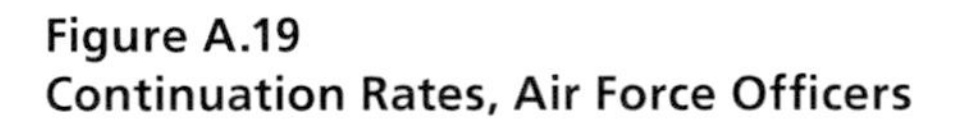

Figure A.19
Continuation Rates, Air Force Officers

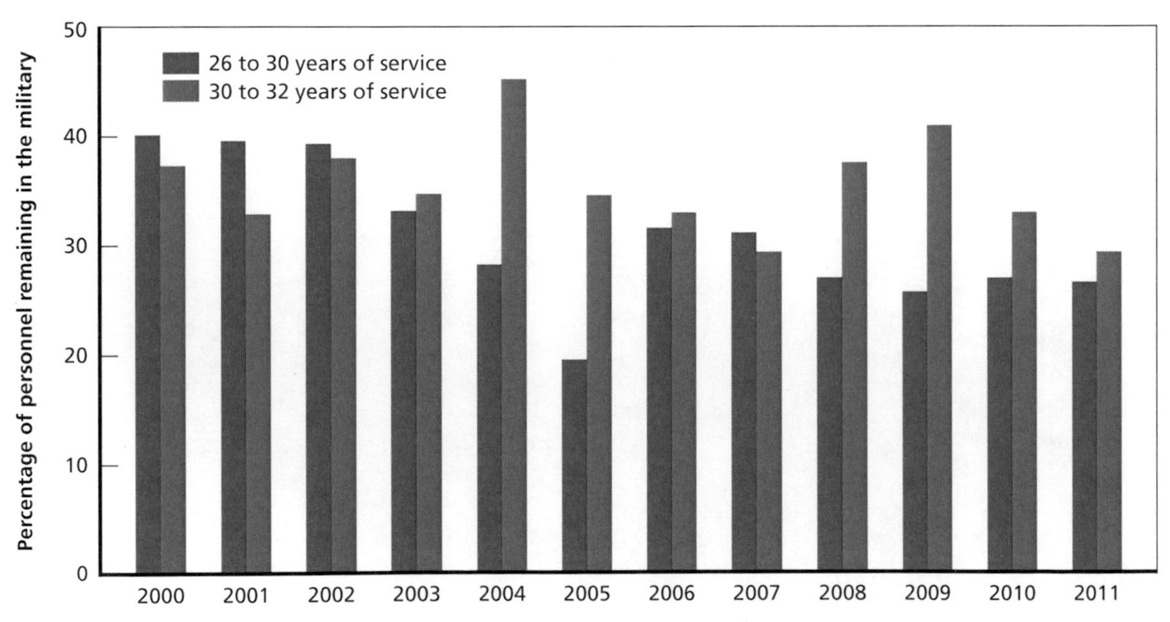

RAND *RR1209-A.19*

Figure A.20
Continuation Rates, Air Force Enlisted Personnel

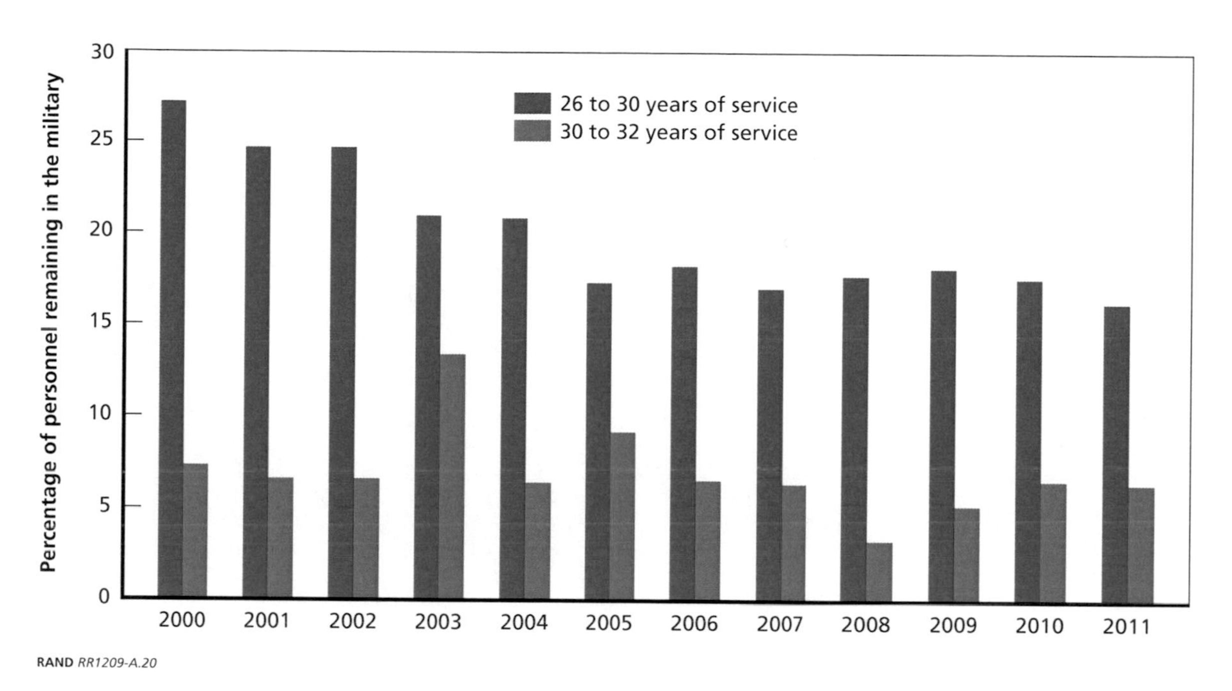

RAND *RR1209-A.20*

Marine Corps Continuation Rates

Figures A.21 through A.23 show continuation rates for Marine Corps officers, warrant officers, and enlisted personnel. For officers, continuation rates for personnel with 26 and 30 years of service have not changed significantly over the period under consideration (Figure A.21). In 2001, 41 percent of officers with 26 years of service remained in the military through YOS 30. This rose to 50 percent in 2002 and stood at 43 percent in 2006. In 2007, the rate of continuation to YOS 30 rose to 54 percent, but this was a temporary increase. The rate fell to 46 percent in 2008 and stood at 46 percent in 2011. The increase in the average continuation rate in the post-2007 period has also been small, from 45 percent before 2007 to 29 percent afterward, and was largely driven by higher-than-normal continuation rates for several years after 2007. Thus, the overall change over the period and when comparing 2006 and subsequent years has been modest, without a clear trend in either direction. For personnel with 30 years of service, there does appear to be some increase in continuation rates to YOS 32, but the upward trend is weakened by several years with much lower continuation rates. The continuation rate for officers with 30 years of service to YOS 32 since 2007 has varied considerably, ranging from 37 to 55 percent. However, this rate has more consistently been above 50 percent since 2005 than in previous years, when continuation rates for these personnel varied between 39 and 49 percent. There is a small increase in the average continuation rate to YOS 32 for personnel with 30 years of service. This suggests a slight increase overall, but one that is modest in size and inconsistent, with several years with very low continuation rates. Furthermore, any upward trend observed seems to begin before 2007.

Looking at warrant officers, continuation rates for personnel with 26 and 30 years of service fluctuate significantly, with no clear upward or downward trend (Figure A.22). Continuation rates for personnel with 26 years of service to YOS 30 range from 32 to 62 percent, while continuation rates for personnel with 30 years of service to YOS 32 range from 6 to 33 percent.

Figure A.21
Continuation Rates, Marine Corps Officers

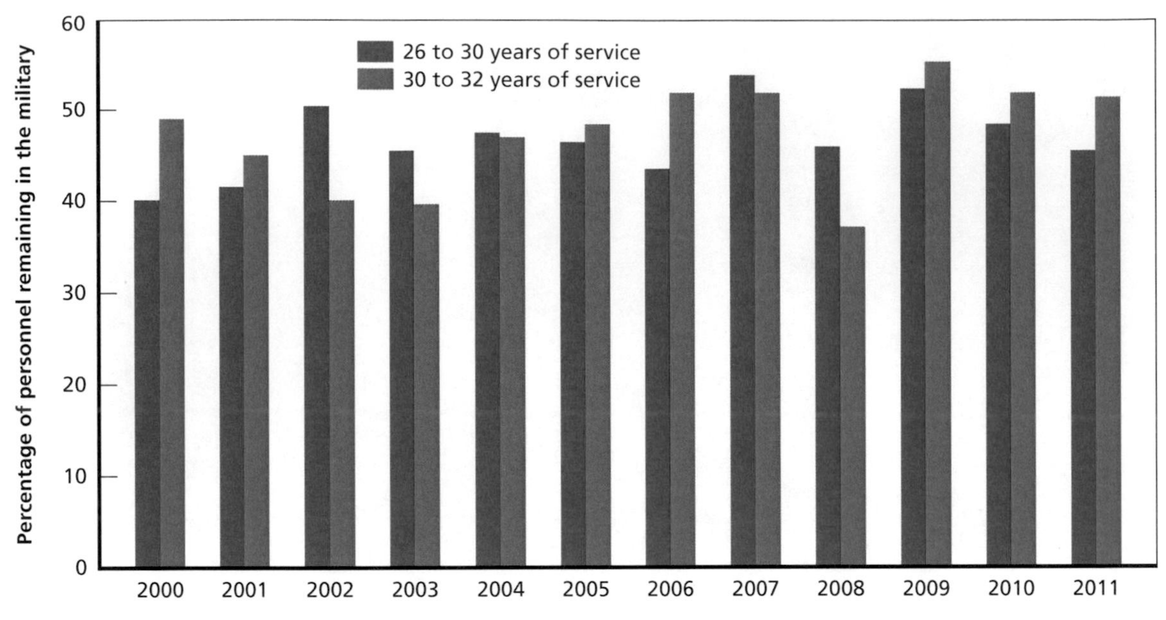

RAND *RR1209-A.21*

Figure A.22
Continuation Rates, Marine Corps Warrant Officers

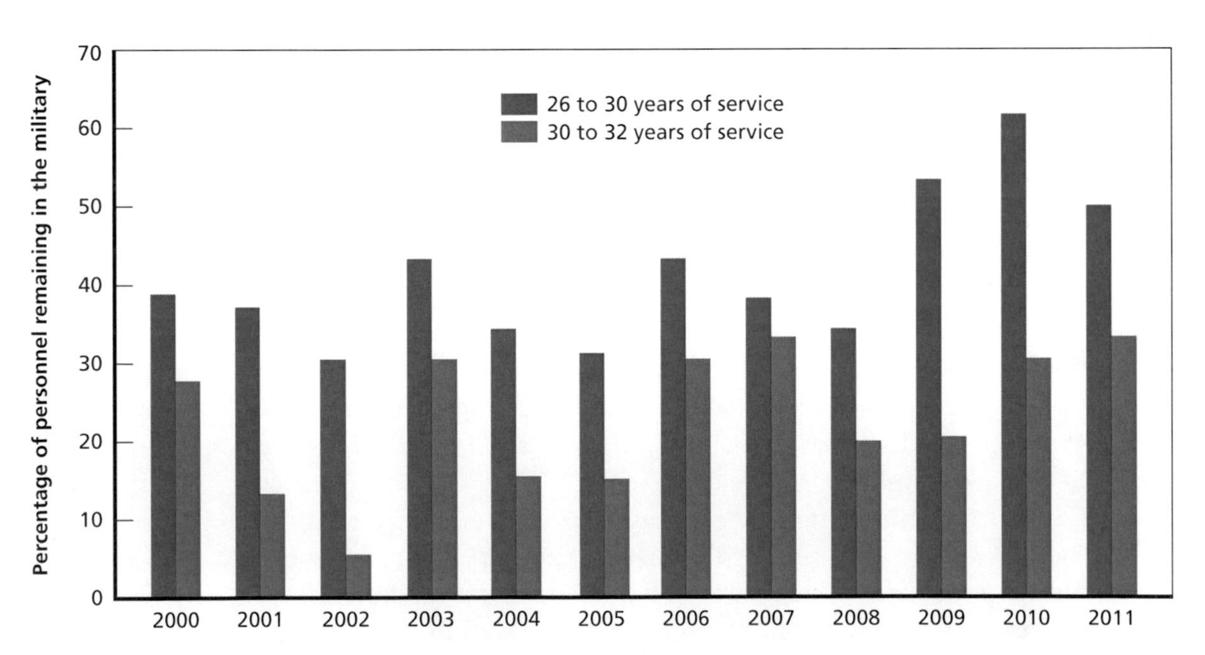

RAND *RR1209-A.22*

It is worth noting that the average continuation rate to YOS 30 for personnel with 26 years of service is higher prior to 2007 than after 2007 (36 percent prior and 47 percent after). The same is true for more-senior warrant officers with 30 years of service continuing to YOS 32 (18 percent prior to 2007 and 28 percent after). This suggests some evidence of an increase in the continuation rates following the 2007 legislative changes for warrant officers with 26 and 30 years of service, despite the lack of clear trends in either case, albeit an increase that is weakened by the considerable variation.

For enlisted Marines, there does appear to be a slight upward trend in the continuation rates of personnel with 26 years of service to YOS 30, both between 2000 and 2011 and since 2007 (Figure A.23). For enlisted personnel with 26 years of service in 2000, the continuation rate to YOS 30 was 35 percent. This rate rose to 44 percent in 2005 and 2006. It then rose further after 2007, from 47 percent in 2007 to 57 percent in 2011. A comparison of the average continuation rate to YOS 30 for personnel with 26 years of service does increase after 2007, from 42 percent prior to 2007 to 51 percent afterward. However, it is worth noting that that this increase is still modest in size and began before the change to the 40-year pay table in 2007. For enlisted personnel with 30 years of service, the continuation rate to YOS 32 has not changed much since 2001, and there has been no sustained upward trend or change in the average continuation rate since 2007. This continuation rate has remained between 15 and 18 percent since 2004.

Overall, continuation rates have not changed significantly for Marine Corps personnel with 26 and 30 years of service. There seems to be a slight upward trend for enlisted personnel with 26 years of service and some evidence of an increase in continuation for warrant officers at 26 and 30 years of service (when looking at average continuation rates). However, even these upward trends are interrupted by considerable variation and years with lower continuation rates. As in the Air Force and Army, then, it does not seem that there has been a marked increase in continuation rates for senior personnel since the change to 40-year pay table in 2007.

Figure A.23
Continuation Rates, Marine Corps Enlisted Personnel

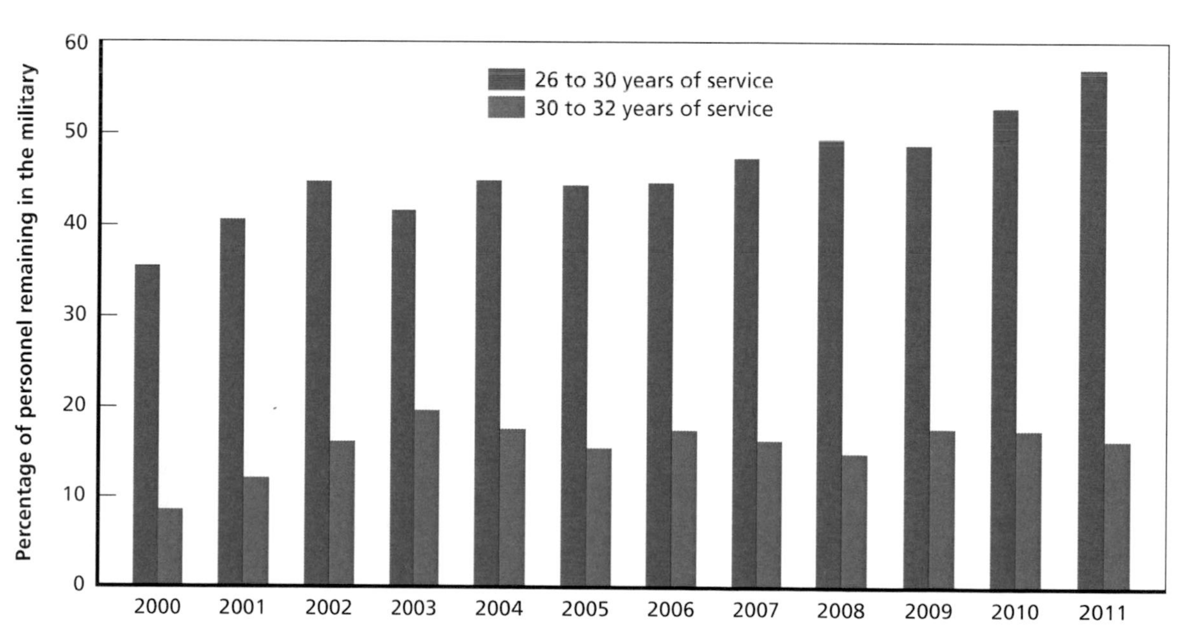

RAND *RR1209-A.23*

Navy Continuation Rates

Figures A.24 through A.26 show the continuation rates for Navy officers, warrant officers, and enlisted personnel with 26 and 30 years of service. As in the other services, there is little evidence of an increase or decrease in continuation rates since 2007. For officers (Figure A.24), there has been almost no variation in continuation rates for personnel with 26 years of service (to YOS 30) or personnel with 30 years of service (to YOS 32). Continuation rates for personnel with 26 years of service to YOS 30 have ranged between 41 and 50 percent, with most years between 45 and 50 percent. For officers with 30 years of service, continuation rates to YOS 32 have ranged primarily from 44 to 55 percent, with two outliers (42 percent in 2000 and 54 percent in 2009). In neither case does the average continuation rate change before and after 2007.

Looking at Navy warrant officers (Figure A.25), there has been considerably more variation but no clear upward or downward trend for personnel with 26 or 30 years of service. For personnel with 26 years of service, continuation rates to YOS 30 range from 46 to 65 percent, although most years, this rate is above 55 percent. There is no difference in the average continuation rate before and after 2007, with continuation rates averaging 57 and 59 percent, respectively. For warrant officers with 30 years of service, continuation rates to YOS 32 range from 5 to 67 percent, although in most years, the rate is between 15 and 26 percent (there are several outliers, including 5 percent in 2005, 13 percent in 2000, and 67 percent in 2009). While the average continuation rate to YOS 32 for personnel with 30 years of service after 2007 is higher than that before 2007, this is likely driven by the outliers (low outliers occur

Figure A.24
Continuation Rates, Navy Officers

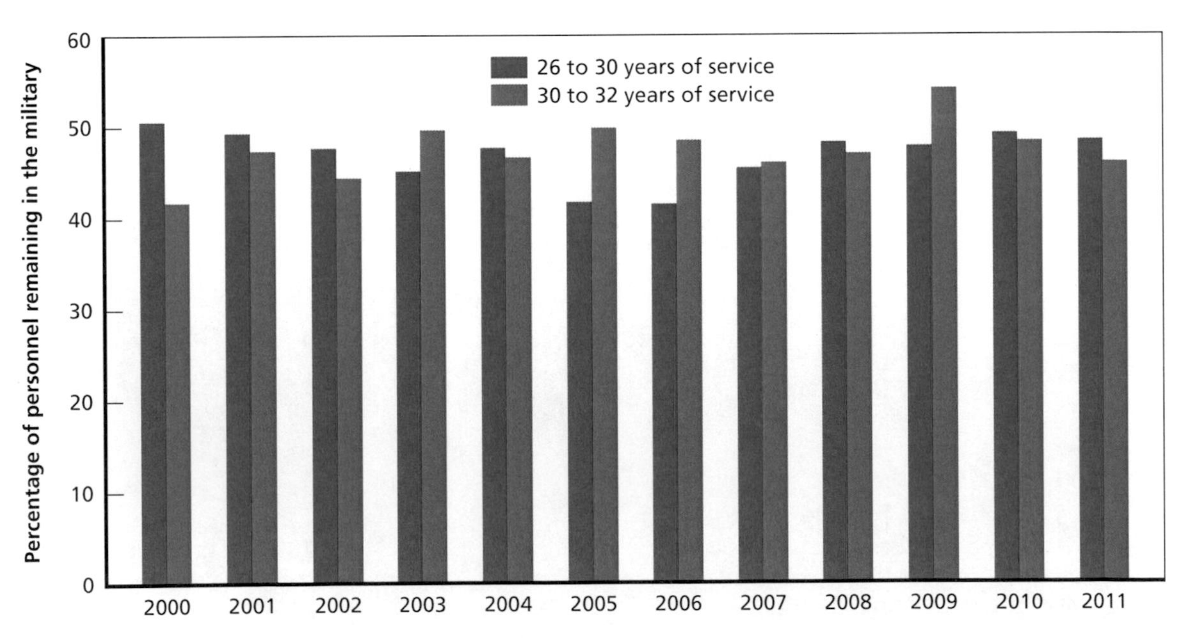

RAND *RR1209-A.24*

Figure A.25
Continuation Rates, Navy Warrant Officers

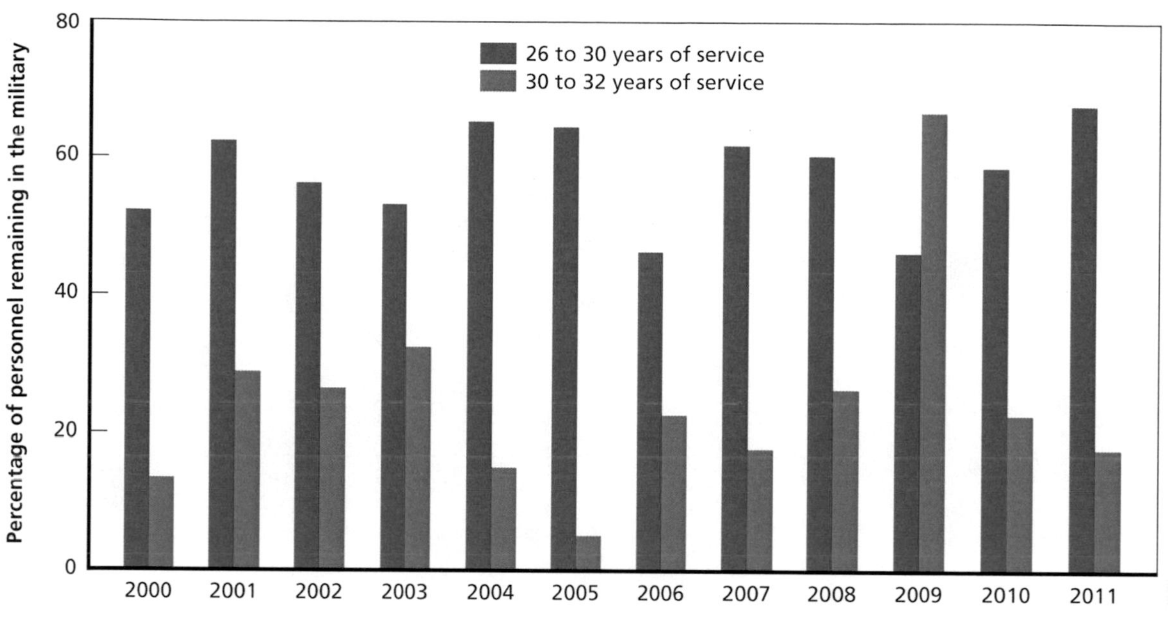

RAND *RR1209-A.25*

before 2007, with higher ones after 2007) and may not reflect a true upward trend in continuation rates for these personnel.

Finally, for enlisted personnel (Figure A.26), continuation rates of personnel with neither 26 years of service nor 30 years of service have changed significantly over the period under consideration or since 2007. For personnel with 26 years of service, continuation rates to YOS 30 have ranged from 27 to 33 percent, with no clear upward or downward trend. Variation since 2007 has been similarly small, with rates between 29 and 33 percent. For enlisted personnel with 30 years of service, continuation rates to YOS 32 have varied from 9 to 16 percent. Since 2007, rates ranged between 9 and 12 percent. Average continuation rates before and after 2007 similarly show no real evidence of an increase or decrease in retention for personnel at 26 or 30 years of service over this period.

Even more than for the other services, there is almost no evidence of a change in continuation rates for Navy personnel, whether looking at the whole period or focusing on the period after 2007.

Figure A.26
Continuation Rates, Navy Enlisted Personnel

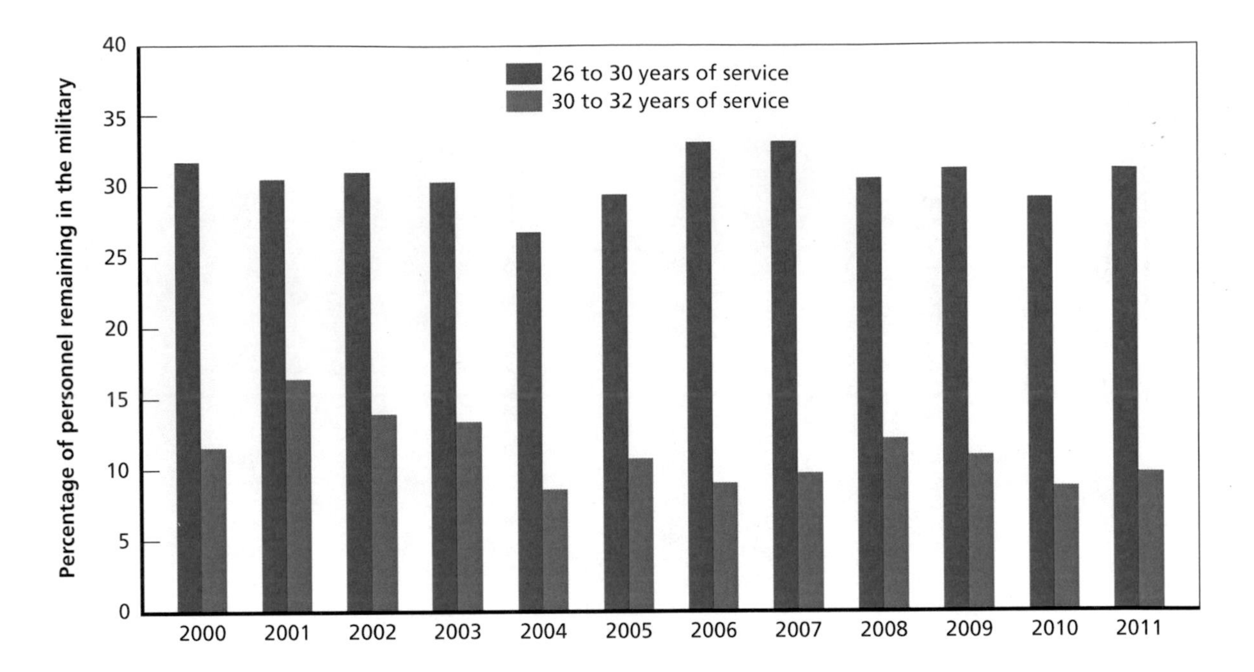

References

Abowd, John, "Does Performance-Based Managerial Compensation Affect Corporate Performance?" *Industrial and Labor Relations Review*, Vol. 43, 1990, pp. 52S–73S.

Asch, Beth J., James Hosek, and Michael G. Mattock, *A Policy Analysis of Reserve Retirement Reform,* Santa Monica, Calif.: RAND Corporation, MG-378-OSD, 2013. As of July 29, 2015:
http://www.rand.org/pubs/monographs/MG378.html

———, *Toward Meaningful Compensation Reform: Research in Support of DoD's Review,* Santa Monica, Calif.: RAND Corporation, RR-501-OSD, 2014. As of July 31, 2015:
http://www.rand.org/pubs/research_reports/RR501.html

Asch, Beth J., Michael G. Mattock, and James Hosek, *A New Tool for Assessing Workforce Management Policies Over Time: Extending the Dynamic Retention Model,* Santa Monica, Calif.: RAND Corporation, RR-113-OSD, 2013. As of July 29, 2015:
http://www.rand.org/pubs/research_reports/RR113

Asch, Beth J., and John T. Warner, *A Theory of Military Compensation and Personnel Policy*, Santa Monica, Calif.: RAND Corporation, MR-439-OSD, 1994. As of July 29, 2015:
http://www.rand.org/pubs/monograph_reports/MR439.html

———, "Should the Military Retirement System Be Reformed?" in J. Eric Fredland, Curtis L. Gilroy, Roger D. Little, and W.S. Sellman, eds., *Professionals on the Front Line: Two Decades of the All-Volunteer Force*, Washington, D.C.: Brassey's, 1996, pp. 175–206.

———, "A Theory of Compensation and Personnel Policy in Hierarchical Organizations with Application to the United States Military," *Journal of Labor Economics*, Vol. 19, No. 3, 2001, pp. 523–562.

Baker, George, Michael Gibbs, and Bengt Holmstrom, "The Internal Economics of the Firm: Evidence from Personnel Data," *Quarterly Journal of Economics*, Vol. 109, 1994, pp. 881–919.

Baker, George, Michael Jensen, and Kevin Murphy, "Compensation and Incentives: Practice vs. Theory," *Journal of Finance*, Vol. 43, 1988, pp. 593–616.

Bognanno, Michael, "Corporate Tournaments," *Journal of Labor Economics*, Vol. 19, No. 2, 2001, pp. 290–315.

Chu, David, "Reform of Basic Pay Rates Legislative Proposal for Fiscal Year 2007," Memorandum for Under Secretary of Defense (Comptroller) and General Counsel of the Department of Defense, March 21, 2006.

Ehrenberg, Ronald, and Michael Bognanno, "Do Tournaments Have Incentive Effects?" *Journal of Political Economy*, Vol. 98, No. 6, 1990, pp. 1307–1324.

Gotz, Glenn A., and John McCall, *A Dynamic Retention Model for Air Force Officers: Theory and Estimates*, Santa Monica, Calif.: RAND Corporation, R-3028-AF, 1984. As of July 29, 2015:
http://www.rand.org/pubs/reports/R3028.html

Harrell, Margaret, Harry Thie, Peter Schirmer, and Kevin Brancato, *Aligning the Stars: Improvements to General and Flag Officer Management,* Santa Monica, Calif.: RAND Corporation, MR-1712-OSD, 2004. As of July 29, 2015:
http://www.rand.org/pubs/monograph_reports/MR1712.html

Henning, Charles, *Military Pay and Benefits: Key Questions and Answers,* Washington, D.C.: Congressional Research Service, 2008.

Lazear, Edward P., and Paul Oyer, "Personnel Economics," in Robert Gibbons and John Roberts, eds., *The Handbook of Organizational Economics*, Princeton, N.J.: Princeton University Press, 2012, pp. 479–519.

Lazear, Edward, and Sherwin Rosen, "Rank-Order Tournaments as Optimum Labor Contracts," *Journal of Political Economy*, Vol. 89, 1981, pp. 841–864.

Leonard, Jonathan, "Executive Pay and Firm Performance," *Industrial and Labor Relations Review*, Vol. 43, 1990, pp. 13S–29S.

Main, Brian, Charles O'Reilly, and James Wade, "Top Executive Pay: Tournaments or Teamwork," *Journal of Labor Economics*, Vol. 11, No. 4, 1993, pp. 606–628.

Malcomson, James, "Work Incentives, Hierarchy, and Internal Labor Markets," *Journal of Political Economy*, Vol. 92, 1984, pp. 486–507.

Mattock, Michael G., James Hosek, and Beth J. Asch, *Reserve Participation and Cost Under a New Approach to Reserve Compensation*, Santa Monica, Calif.: RAND Corporation, MG-1153-OSD, 2012. As of July 29, 2015: http://www.rand.org/pubs/monographs/MG1153.html

Office of the Secretary of Defense, *Proposal for a Reform of Basic Pay Rates for FY 2007: Section-by-Section Analysis*, Washington, D.C.: U.S. Department of Defense, April 1, 2006.

OSD—*See* Office of the Secretary of Defense.

Philpott, Tom, "Congress: We Went Too Far on Star-Rank Retirement," *Stars and Stripes*, December 31, 2014.

Pleeter, Saul, *Pros and Cons of Reform of Basic Pay Rates for Fiscal Year 2007*, Washington, D.C.: U.S. Department of Defense, Office of the Secretary of Defense of Personnel and Readiness, white paper, March 21, 2006.

Prendergast, Canice, "The Provision of Incentives in Firms," *The Journal of Economic Literature*, Vol. 37, No. 1, 1999, pp. 7–63.

Public Law 109-364, John Warner National Defense Authorization Act for Fiscal Year 2007, October 17, 2006.

Public Law 113-291, Carl Levin and Howard P. "Buck" McKeon National Defense Authorization Act for Fiscal Year 2015, December 2014.

Rosen, Sherwin, "Prizes and Incentives in Elimination Tournaments," *American Economic Review*, Vol. 76, 1986, pp. 701–715.

———, "The Military as an Internal Labor Market: Some Allocative, Productivity, and Incentive Problems," *Social Science Quarterly*, Vol. 73, 1992, pp. 49–64.

SASC—*See* Senate Armed Services Committee.

Senate Armed Services Committee, S. Rept. 113-176, 113th Congress, June 4, 2014.

Vanden Brook, Tom, "Retired Pay for O-10s Spiked with Rule Changes," *USA Today,* February 27, 2012.

Willis, Robert, and Sherwin Rosen, "Education and Self-Selection," *Journal of Political Economy*, Vol. 87, No. 5, 1979, pp. S7–S36.